Revalidating Process Hazard Analyses

This book is one of a series of titles published by the Center for Chemical Process Safety. A complete list of titles available appears at the end of this book.

Revalidating Process Hazard Analyses

Walter L. Frank and David K. Whittle
EQE International, Inc.

CENTER FOR CHEMICAL PROCESS SAFETY
American Institute of Chemical Engineers
3 Park Avenue, New York, NY 10016-5991

Library of Congress Cataloging-in-Publication Data
Frank, Walter L.
Revalidating process hazard analyses / Walter L. Frank and David K. Whittle.
p. cm.
Includes bibliographical references and index.
ISBN 0-8169-0830-3
1. Chemical processes—Safety measures. I. Whittle, David K. II. Title.
TP150 .S24 F7 2000
660'.2804--dc21
00-051103

Contents

Preface

The American Institute of Chemical Engineers (AIChE) has a 30-year history of involvement with process safety for chemical processing plants. Through its strong ties with process designers, builders, operators, safety professionals and academia the AIChE has enhanced communication and fostered improvement in the high safety standards of the industry. AIChE publications and symposia have become an information resource for the chemical engineering profession on the causes of accidents and means of prevention.

The Center for Chemical Process Safety (CCPS), a directorate of AIChE, was established in 1985 to develop and disseminate technical information for use in the prevention of major chemical accidents. The CCPS is supported by a diverse group of industrial sponsors in the chemical process industry and related industries who provide the necessary funding and professional guidance for its projects. The CCPS Technical Steering Committee selects the projects to be developed, with approval of the Advisory Board, and oversees the individual projects selected.

In 1992, CCPS published *Guidelines for Hazard Evaluation Procedures, Second Edition, with Worked Examples* to provide information to organizations for conducting hazard evaluations of processes handling potentially hazardous materials. While many companies had previously adopted the practice of reviewing process hazard analyses on periodic basis, there are now regulatory requirements mandating such revalidations for certain processes.

The goal of this guideline is to provide plant management, operating personnel, engineering groups, and safety professionals with supplemental information and methods to achieve well-executed revalidations, recognizing that there are many alternate routes to achieving a high level of quality in process hazard reviews.

Acknowledgments

The American Institute of Chemical Engineers (AIChE) wishes to thank the Center for Chemical Process Safety (CCPS) and those involved in its operation, including its many sponsors whose funding made this project possible, and the members of the Technical Steering Committee who conceived of and supported this CCPS Concept project.

This project was initiated as a Workshop at the September 1999 Technical Steering Committee (TSC) meeting in San Francisco. We wish to recognize the contributions of the TSC members and the Workshop speakers:

Jeff M. Gunderson, Chevron Corporation
David W. Jones, EQE International, Inc.
Mike Marshall, DOL-OSHA
Jon F. Plakosh, ATOFINA Chemicals, Inc.

In addition, W. C. (Bill) Geckler of EQE International, Inc. ably assisted in facilitating the Workshop break-out sessions.

The members of the PHA Revalidation Subcommittee who worked with EQE International, Inc. to produce this text deserve special recognition for their dedicated efforts, technical contributions, and overall enthusiasm for creating a useful addition to the CCPS Concept series.

The members of the Subcommittee were:

Dennis Blowers, Solvay Polymers, Inc.
Donald J. Connolley, Akzo Nobel Chemicals Inc.
Peter McGrath, Olin Corporation
Jon F. Plakosh, ATOFINA Chemicals, Inc.
Marty Welch, Chevron Corporation

Ray E. Witter was the CCPS staff liaison and was responsible for the overall administration of the project.

EQE International, Inc. was the contractor for this project and Walter L. Frank and David K. Whittle were the principal authors of the text. The authors relied significantly upon and wish to acknowledge the prior work, thoughtful review, and support of co-workers David Jones, Steve Arendt, Kevin Smith, Jack Vernon, and Charles Foshee.

CCPS also gratefully acknowledges the comments and the suggestions submitted by the following peer reviewers:

Al W. Bickum, The Goodyear Tire & Rubber Company
Michael P. Broadribb, BP Amoco
Laurie J. Brown, Eastman Chemical Company
C. Curtis Clements, DuPont Company
John DiPalma, CYTEC Industries
Carol N. Garland, Eastman Chemical Company
Harry J. Glidden, DuPont Company
Dr. Martin Gluckstein, CCPS Staff
Rashid Hamsayeh, Formosa Plastics Corporation (USA)
Linda Hicks, Reilly Industries, Inc.
Russell Kahn, Novartis Corporation
Peter N. Lodal, Eastman Chemical Company
Lisa Morrison, Nova Chemicals
David J. Repasky, Sherwin Williams Company
Robert M. Rosen, BASF Corporation
Edward J. Ryczek, Merck & Company, Inc.
Adrian L. Sepeda, Occidental Chemical Corp.
Ronnie Tucker, PPG Industries, Inc.

Their insights, comments, and suggestions helped ensure a balanced perspective for this Concept Series book.

Lastly, we wish to express our appreciation to Jack Weaver and Les Wittenberg of the CCPS staff for their support and guidance.

Glossary

Alternative Release Scenario (ARS) The basis for an off-site consequence analysis required by the EPA RMP Rule. This release scenario is less conservative, and more likely to occur than the Worst Case Scenario.

Consequences The direct, undesirable result of an accident sequence usually involving a fire, explosion, or release of toxic material. Consequence descriptions may be qualitative or quantitative estimates of the effects of an accident.

Covered Process A process subject to regulatory requirements established under the OSHA PSM Standard or the EPA RMP Rule.

Essential Criteria Criteria defining the required content or conduct of a PHA based upon company or regulatory requirements.

Hazard An inherent physical or chemical characteristic that has the potential for causing harm to people, property, or the environment.

Hazard Evaluation (HE) The analysis of the significance of hazardous situations associated with a process or activity. Uses qualitative techniques to pinpoint weaknesses in the design and operation of facilities that could lead to accidents.

Human Factors A discipline concerned with designing machines, operations, and work environments to match human capabilities, limitations, and needs.

Incident Investigation The management process by which underlying causes of undesirable events are uncovered and necessary steps are taken to prevent similar occurrences.

Likelihood A measure of the expected probability or frequency of an event's occurrence.

Management Of Change (MOC) A system to identify, review and approve all modifications to equipment, procedures, raw materials and processing conditions, other than "replacement in kind," prior to implementation.

Node Sections of equipment with definite boundaries (e.g., a line between two vessels) within which process parameters are investigated for deviations. The locations on P&IDs at which the process parameters are investigated for deviations (e.g., a reactor). The concept of dividing a process into nodes for analysis is commonly, but not exclusively, used in HAZOPs.

Pre-Startup Safety Review (PSSR) A system to confirm, before the introduction of hazardous chemicals, that new or modified facilities are in accordance with design specifications; adequate procedures are in place; appropriate hazard analyses or management of change reviews have been conducted; and training of affected personnel has been completed.

Process Hazard Analyses (PHA) An organized effort to identify and evaluate hazards associated with chemical processes and operations to enable their control. This review normally involves the use of qualitative techniques to identify and assess the significance of hazards. Conclusions and appropriate recommendations are developed. Occasionally, quantitative methods are used to help prioritize risk reduction efforts.

Process Safety Information (PSI) Physical, chemical, and toxicological information related to the chemicals, process, and equipment. It is used to document the configuration of a process, its characteristics, its limitations, and as data for process hazard analyses.

Process Safety Management (PSM) A program or activity involving the application of management principles and analytical techniques to ensure the safety of chemical process facilities. Sometimes called process hazard management. Each principle is often termed an "element" or "component" of process safety.

Program Level 1 An implementation program level under the EPA RMP Rule. Applies to those facilities perceived to have a negligible potential for serious off-site consequences in the event of an accidental release. There are no requirements for an accidental release prevention program for this program level.

Program Level 2 An implementation program level under the EPA RMP Rule. Applies to certain facilities perceived to have the potential for off-site consequences in the event of an accidental release. The accidental release prevention program requirements for this program level are less rigorous than the Program Level 3 requirements.

Program Level 3 An implementation program level under the EPA RMP Rule. Applies to certain facilities perceived to have the potential for off-site consequences in the event of an accidental release. The accidental release prevention program for this program level contains the most rigorous and detailed requirements under the RMP Rule.

Recommendation A suggested course of action intended to prevent the occurrence (or recurrence) of an accident event sequence, or to mitigate its consequences.

Redo To conduct a new PHA.

Retrofit, Update, and Revalidate To correct substantive deficiencies in the prior PHA, and then update and revalidate it (See Update and Revalidate).

Risk Management Program (RMP) Rule EPA's accidental release prevention Rule, which requires covered facilities to prepare, submit, and implement a risk management plan.

Update and Revalidate To revise a prior PHA, as required, to reflect any changes that have occurred since the prior PHA; new learnings about the hazards of the process; changes in risk management requirements; etc.

Worst Case Scenario (WCS) The basis for an off-site consequence analysis required by the EPA RMP Rule. This intentionally conservative accident scenario assumes the release of the entire inventory of a vessel, under the most unfavorable conditions, and with the failure of most protective features.

Acronyms and Abbreviations

ACC	American Chemical Council; formerly the Chemical Manufacturers Association (CMA)
AIChE	American Institute of Chemical Engineers
API	American Petroleum Institute
ARS	Alternative Release Scenario
CAA	Clean Air Act
CCPS	Center for Chemical Process Safety
CMA	Chemical Manufacturers Association (See ACC)
CPI	Chemical Process Industry
DCS	Distributed Control System
EPA	Environmental Protection Agency
FMEA	Failure Mode and Effects Analysis
FTA	Fault Tree Analysis
HAZOP	HAZard And OPerability (Study)
HE	Hazard Evaluation
MOC	Management of Change
OSHA	Occupational Safety and Health Administration
PHA	Process Hazard Analysis
PSI	Process Safety Information
PSM	Process Safety Management
PSSR	Pre-Startup Safety Review
RMP	Risk Management Plan
WCS	Worst Case Scenario

Introduction

Why Was This Book Written?

Formal Process Hazard Analyses (PHAs) have been performed in the chemical process industries (CPI) for nearly 40 years, and most companies have included requirements for the conduct of PHAs in their company process safety programs. CCPS and industry organizations such as the American Petroleum Institute (API) and the American Chemical Council (ACC, formerly the Chemical Manufacturers Association) have followed industry's lead and have incorporated PHA as one of the elements in model process safety management (PSM) programs prepared for their member companies.

Many organizations include a periodic update requirement in their PHA programs. Such updates are intended to address considerations such as changes in processes and procedures, or increased knowledge about the hazards of the process. For example, CCPS in its *Plant Guidelines for Technical Management of Chemical Process Safety* provides guidance for process risk management assessments, which include a PHA. CCPS states:

> It is suggested that the frequency of a process risk management assessment should be every 3 to 5 years... The frequency... may also be influenced by incidents or accidents, process changes, etc.

API in its Recommended Practice 750, *Management of Process Hazards*, suggests that (API 1990):

> PHAs should be reviewed and updated periodically, with typical review intervals ranging from 3 to 10 years. ... changes in technology or in the facility... should be considered in establishing the review frequency...

Finally, ACC, in the Technology section of its Process Safety Code of Management Practices, suggests the *"Periodic assessment and documentation of process hazards..."* (ACC 1990).

Working with considerable input from industry, OSHA promulgated its Process Safety Management (PSM) Standard (29 CFR 1910.119) in February

1992 and EPA followed with its Risk Management Program (RMP) Rule (40 CFR Part 68) in June 1996 (OSHA 1992, EPA 1996). Both regulations require that regulated facilities conduct PHAs for covered processes and that PHAs be updated and revalidated at an interval not to exceed five years.

A properly conducted PHA can involve a significant investment of time and organizational resources. Thus, it is important that the PHA effort (both initial and revalidation) be wisely directed and implemented to properly steward limited resources. A considerable body of literature has been written on procedures and practices for conducting PHAs. The CCPS publication, *Guidelines for Hazard Evaluation Procedures, Second Edition, with Worked Examples* (subsequently referred to in this book as the *HEP Guidelines*), provides detailed guidance on topics such as preparing for the study, application of the various techniques, and analysis follow-up considerations (CCPS G-1).

This book addresses the need, identified by CCPS, for supplemental guidance with respect to considerations unique to the PHA revalidation task.

Scope of This Book

This book provides one approach, and a number of tools, which should be applicable in a variety of situations. However, it does not provide a "one-size-fits-all" solution to the revalidation task. Neither is it intended to provide the "one true solution" to the task. Readers should develop procedures based upon an evaluation of their specific requirements.

This book assumes that the reader has at least a basic familiarity with the conduct of a PHA. While some introductory material is provided in Chapter 1, this book is not intended as an instructional text covering the general conduct of a PHA. Readers needing such detail are referred to the *HEP Guidelines.*

This book is a supplement to the *HEP Guidelines* and is premised on the same concept stated in this earlier work:

> This book does not contain a complete program for managing the risk of chemical operations, nor does it give specific advice on how to establish a hazard analysis program for a facility or an organization. However, it does provide some insights that should be considered when making risk management decisions and designing risk management programs.

This book outlines a demonstrated, commonsense approach for resource-effective PHA revalidation. This approach first examines a number of factors such as the quality of the prior PHA; the frequency and magnitude of changes that have occurred within the process; the adequacy of the prior

PHA documentation; and the frequency of incidents and near-misses within the process. A revalidation plan is developed based upon this input.

This book is intended to be a concise reference for both the plant PHA practitioner and those with management responsibility for PSM implementation. Flowcharts, checklists, and worksheets provide the reader with valuable tools for use in the revalidation process.

The CCPS Workshop

CCPS, at the September 1999 Technical Steering Committee meeting in San Francisco, made PHA Revalidation the topic of an afternoon workshop. This workshop included breakout sessions where participants had a chance to share their thoughts and experiences on the topic. The breakout groups addressed three basic questions:

1. What tasks should be completed before or during the team reviews of the PHA under consideration?
2. What criteria should be used to decide whether a PHA is valid in its current form, portions of the PHA must be revised, or the entire PHA needs to be redone?
3. What learnings can be applied to making future revalidations easier?

The work product from the workshop has been incorporated into this book.

How This Book Is Organized

Chapter 1 provides an introductory overview of process hazard analysis and what it is intended to accomplish. A brief synopsis of the more common PHA techniques is provided.

Chapter 2 provides a general background on PHA revalidation: What is it? Why is it important? When should it be done?

Chapter 3 discusses the efforts involved in preparing for the revalidation task. Topics include preplanning the revalidation, and identifying, collecting, and preparing the information needed to support the task.

Chapter 4 provides instructions for evaluating the completeness and quality of the prior PHA.

Chapter 5 provides background and detailed instructions for identifying the changes that have been introduced into the process, equipment, and facilities since the prior PHA was completed. These changes will be considered during the revalidation.

Chapter 6 describes the alternative approaches available for use in the PHA revalidation and provides guidance on how to identify the appropriate revalidation methodology.

Chapter 7 provides guidance on conducting the PHA revalidation study sessions.

Chapter 8 addresses the documentation of the PHA revalidation study.

Finally, Appendices are provided containing supporting information and valuable checklists for use in conducting PHA revalidations.

I

Refresher on the Basics

Before getting into the specifics of process hazard analysis (PHA) revalidation, it might be helpful to briefly review the intent and conduct of PHAs.

1.1. What a PHA Is Intended to Accomplish

In 1992, CCPS updated and republished its *Guidelines for Hazard Evaluation Procedures.* In this second edition, CCPS stated:

> A hazard is an inherent physical or chemical characteristic that has the potential to do harm (to people, property, or the environment). A hazard evaluation (HE) study is an organized effort to identify and analyze the significance of hazardous situations associated with a process or activity. . . . [It is] used to pinpoint weaknesses in the design and operation of facilities that could lead to accidental chemical releases, fires, or explosions. . . . [It provides] organizations with the information to help them improve the safety and manage the risk of their operations.

As shown in Figure 1.1, the PHA process involves five steps:

1. Identify the hazardous characteristics within the process. These may be characteristics of the process materials (e.g., toxicity or flammability), characteristics of the process (e.g., high temperature or high pressure), or characteristics of the equipment;
2. Identify potential equipment failures or human errors that could lead to accident scenarios that might result in harm;
3. Evaluate the magnitude of the harm (consequences) that could result from the accident scenarios and the likelihood of the accident scenarios occurring, then determine the associated risk based upon the likelihood and consequence;
4. Decide whether that level of risk is tolerable; and
5. For high-risk scenarios, propose risk reduction alternatives (recommendations) to reduce risk to a tolerable level.

HEP...PHA??

CCPS uses the term "hazard evaluation procedures" to encompass a broad range of applications for the analytical tools described in the Guideline. Process Hazard Analysis is one such specific application.

Thus, a PHA is a proactive exercise intended to identify the safety-related weaknesses in a process, its equipment design, its operating practices and procedures, its maintenance practices, etc. Recommended upgrades are proposed to address those weak spots that pose an intolerable risk to personnel, property, or the environment.

1.2. Brief Review of the More Common PHA Methodologies

While there are many factors that go into making a good PHA, proper selection of methodology is one of the most basic. In practical application, six techniques are more commonly used:

1. What-If;
2. Checklist;
3. What-If/Checklist;
4. Hazard and Operability Study (HAZOP);
5. Failure Mode and Effects Analysis (FMEA); and
6. Fault Tree Analysis (FTA).

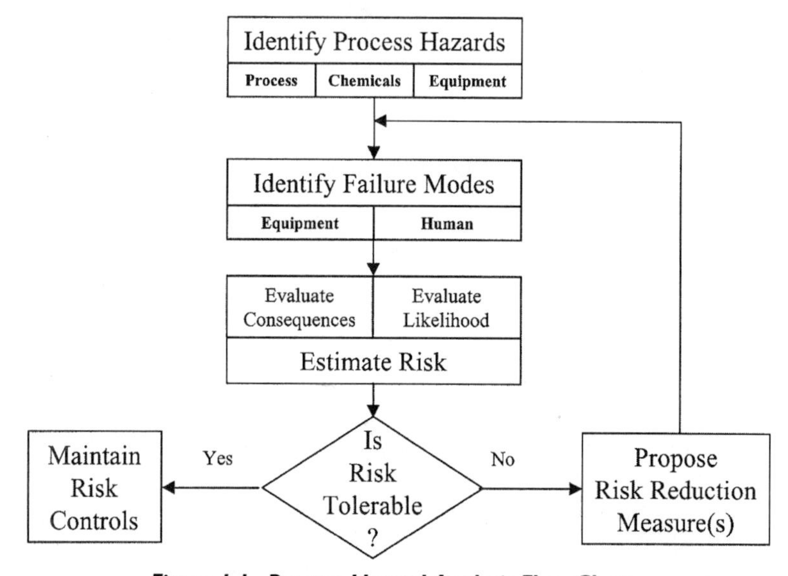

Figure 1.1. Process Hazard Analysis Flow Chart.

A detailed discussion and comparison of the various PHA methodologies is beyond the scope of this book; readers unfamiliar with the various methodologies may wish to refer to the CCPS *HEP Guidelines* for guidance on technique selection. For our purposes here, short descriptions of these methodologies have been abridged from the *HEP Guidelines*, and are provided below.

What-If. The What-If Analysis technique is a brainstorming approach in which team members familiar with the subject process ask questions or voice concerns about possible undesired events. The purpose of a What-If Analysis is to identify hazards, hazardous situations, or specific accident events that could produce an undesirable consequence. An experienced group of people identifies possible accident situations, their consequences, and existing safeguards, then suggests alternatives for risk reduction. The method can involve examination of possible deviations from the design, construction, modification, or operating intent. The What If Analysis requires a basic understanding of the process intention, along with the ability to mentally combine possible deviations from the design intent that could result in an accident.

The concerns (often phrased in the form of *what if?* questions) are formulated based on experience and applied to existing drawings and process descriptions for an operating plant. There may be no specific pattern or order to these questions, unless the leader provides a logical pattern such as dividing the process into functional systems. The questions can address any abnormal condition related to the plant, not just component failures or process variations.

Checklist. A Checklist Analysis uses a written list of items to verify the status of a system. Frequently, checklists are created by simply organizing information from current relevant codes, standards, and regulations. A checklist is typically completed with "yes," "no," "not applicable," or "needs more information" answers to the questions. Qualitative results vary with the specific situation, but generally, they lead to a decision about compliance with standard procedures and recognized practices. In addition, knowledge of any discovered deficiencies allows development of a list of possible safety improvement recommendations for consideration. The Checklist Analysis method is versatile. The type of evaluation performed with a checklist can vary. It can be used quickly for simple evaluations or for more extensive in-depth evaluations.

What-If/Checklist. The What-If/Checklist Analysis technique combines the creative, brainstorming features of the What-If Analysis method with the systematic features of the Checklist Analysis method. The hybrid method capitalizes on the strengths and compensates for the individual shortcomings of the separate approaches. For example, the Checklist Analysis method is an experience-based technique, and the quality of a PHA performed using this approach is highly dependent on the experience of the checklist's authors. If the checklist is not complete, then the analysis may not effectively capture a hazardous situation. The What-If Analysis portion of the technique encourages the team to consider potential accident events and consequences that are beyond the experience of the checklist authors and, thus, are not covered on the checklist. Conversely, the checklist portion of this technique lends a more systematic nature to the What-If Analysis.

Hazard and Operability Study (HAZOP). The purpose of a HAZOP is to carefully review a process or operation in a systematic fashion to determine whether process deviations can lead to undesirable consequences. In a HAZOP, the team uses a creative, systematic approach to identify hazard and operability problems resulting from deviations from the design intent of the process that could lead to undesirable consequences. An experienced team leader systematically guides the team through the plant design using a fixed set of words (called "guide words"). These guide words are applied at specific points or "study nodes" in the plant design and are combined with specific process parameters to identify potential deviations from the plant's intended operation. For example, the guide word "No" combined with the process parameter "flow" results in the deviation "No Flow."

The HAZOP team lists potential causes and consequences of the deviation as well as existing safeguards protecting against the deviation. When the team determines that inadequate protection exists for a credible deviation, it usually recommends that action be taken to reduce the risk.

Failure Mode and Effects Analysis (FMEA). The purpose of an FMEA is to identify and tabulate equipment and system failure modes and their effects on a system or plant. The failure mode describes how equipment fails (open, closed, on, off, leaks, etc.) The effect of the failure mode is determined by the system's response to the equipment failure. An FMEA identifies single failure modes that either directly result in or contribute significantly to an accident. An FMEA generates a qualitative, systematic reference list of equipment, failure modes, and effects. This analysis typically generates recommendations for increasing equipment reliability, thus improving process safety.

Fault Tree Analysis. Fault Tree Analysis (FTA) is a deductive technique that focuses on one particular accident or main system failure, and provides a method for determining causes of that event. The fault tree is a graphical model that displays the various combinations of equipment failures and human errors that can result in the main system failure of interest (called the Top Event). This allows the hazard analyst to focus preventive or mitigative measures on significant basic causes to reduce the likelihood of an accident.

FTA is well suited for analyses of highly redundant systems. For systems particularly vulnerable to single failures that can lead to accidents, it is better to use a single-failure-oriented technique such as What If, or HAZOP. FTA is often employed in situations where another PHA technique (e.g., HAZOP) has pinpointed an important accident of interest that requires analysis that is more detailed.

1.3. PHA Team Make-up

Analysis team composition is a critical part of any PHA. Industry experience has shown that PHAs are most effectively conducted by a multi-disciplinary team of experienced personnel (Frank 1993). A successful team includes personnel with knowledge and experience specific to the process being analyzed to help:

- Identify hazards that are specific to the unit;
- Provide insight into how the unit responds during upset conditions;
- Relate actual operating, maintenance, and training practices;
- Determine reasonable worst-case effects from upsets; and
- Judge the adequacy of existing safeguards.

Without adequate knowledge and experience specific to the process, a PHA team may not identify unique hazards or appropriately assess potential causes, consequences, or the effectiveness of safeguards to control process hazards.

One of the most important team member selections is that of the team leader. Skilled leadership by someone practiced in the PHA technique to be applied is important to ensure the success of the PHA effort (see sidebar). Participation by someone with an engineering background to help address technical aspects of the PHA deliberations is also important. Guidance on the selection of PHA teams is provided in the *HEP Guidelines.*

Revalidation Study leader should be:

- Knowledgeable in PHA methodologies used;
- A thorough, effective planner;
- Familiar with relevant PSM elements, especially MOC;
- An excellent communicator with good interpersonal skills

2

Revalidation—What Is It?

This chapter describes what a PHA revalidation is intended to accomplish and introduces some of the optional approaches available to consider in the PHA revalidation process.

2.1. The Reason for Revalidation

There are at least six basic reasons why it is important to update and revalidate a PHA on a periodic basis:

re·val·i·date (rē-văl-ĭ-dāt′) *verb, transitive* To declare valid again.

American Heritage® Dictionary of the English Language, Third Edition ©1996, Houghton Mifflin Co.

Changes in Process or Equipment. Process and equipment changes must be anticipated, and most process safety management system models embody a Management of Change (MOC) element to ensure that changes are properly scrutinized and authorized prior to implementation. However, MOC requests are often evaluated within a rather narrow context; i.e., "What are the potential consequences of what I intend to do right now?" There is a risk that, for processes undergoing frequent change, the significance of a particular change may not be assessed within the context of all the other changes that have occurred since the last PHA. Further, not all changes may be captured and evaluated under the MOC program. Thus, revalidation offers an opportunity, on a periodic basis, to perform an integrated evaluation of the cumulative (and potentially synergistic) impact of all of these changes, both controlled and uncontrolled.

Gaps and Deficiencies in Prior PHA. Gaps are errors of omission; i.e., failures to address the established requirements for the conduct of the PHA. Some companies have documented requirements for what a PHA must address. Also, the OSHA and EPA regulations, as will be discussed more fully later, are very specific in defining considerations that must be addressed during PHAs for covered processes; e.g., the consideration of human factors. Failure to address human factors, or some other requisite company or regulatory consideration, would be a gap that must be filled during the PHA revalidation.

Deficiencies, on the other hand, are errors in applying the PHA methodology. For example, the prior PHA Team may not have consistently traced the effects of failure scenarios to their ultimate consequences. Revalidation teams, often including several new members (or, conceivably constituted with all new members) may identify hazards that the prior team overlooked. Even if the same team is used, new learnings may be discovered during the revalidation due to the experience gained by the team members in the interim.

New Knowledge. PHA revalidation teams may have access to information that the prior PHA team members did not have available to them. Such information might come from new company research, from work done by others and reported in the industry literature, or from learnings from incident investigations. New information may result in new or different conclusions or recommendations in the PHA revalidation report. Each revalidation offers a new opportunity to reconsider past deliberations in the light of current knowledge.

Unresolved Recommendations. PHA recommendations should normally be resolved in a timely manner to ensure that protections recommended by the PHA are promptly implemented. However, extenuating circumstances can dictate a delay in implementing a recommendation. The revalidation process provides an opportunity to reevaluate the validity of previous recommendations. If there are real safety issues at stake, why hasn't the recommendation been adopted within the period between the prior PHA and the revalidation PHA? On the other hand, if there is no compelling safety urgency associated with the recommendation, why not close it out as unneeded?

On a more positive note, the revalidation process also affords the team an opportunity to affirm that the actions taken to implement a recommendation did, indeed, resolve the issue addressed by the recommendation.

Regulatory Requirements. Those facilities subject to either the OSHA PSM or EPA RMP regulations must satisfy the regulatory requirement that PHAs be updated and revalidated at least every five years. The relevant portions of the OSHA and EPA regulations are reproduced in Appendix A.

Changing Requirements. The need to update a PHA may be driven by changes in requirements established by internal or external authorities. An example of the prior could be the revision of the company policy or procedure for the conduct of PHAs.

2.2. Revalidation Objective

The primary objective of a PHA revalidation is to produce an updated PHA that adequately identifies, evaluates, and proposes controls for the hazards of the process, *as they are currently understood*. There are a number of reasons why today's understanding of the hazards associated with the process might differ from the understanding that existed at the time of the prior PHA. These could include:

- Process changes have introduced new hazards or accentuated existing hazards.
- Changes in on-site or off-site occupancy patterns have changed the at-risk populations.
- New knowledge is now available to better understand the hazard potential, revealing potentially more severe consequences.
- Actual incidents have revealed scenarios not previously identified in the PHA.
- Safeguards previously credited in the PHA have been removed, compromised, or discredited.

While the factors cited above have been selected to illustrate potentially negative outcomes, positive changes that should be reflected in the PHA revalidation are also possible. Perhaps process changes have actually removed hazards from the process (e.g., through application of inherently safer design principles). Safeguards may exist for which proper credit was not taken in the prior PHA or other new safeguards may have been added in the interim. In such circumstances, the PHA can be modified to reflect the current, less severe risk perspective.

A secondary objective of the revalidation effort might be to accomplish the primary objective with the most efficient use of time, personnel, and resources. Workforces are generally getting smaller, often with more to do, and it is important that the revalidation effort be accomplished with maximum effectiveness.

2.3. Revalidation Concept

The revalidation concept is straightforward:

$$\text{Prior PHA(s)} + \text{Update} \xrightarrow{\text{Revalidation}} \text{New PHA}$$

Considerable time, effort, and thought likely went into conducting the prior PHA. The revalidation process attempts to protect this investment by identifying and building upon the still pertinent portions of the prior PHA. Corrections are made and new content is added to the PHA as required (i.e., the PHA is *Updated*), and the results are documented to serve as the new PHA (i.e., the PHA is *Revalidated*).

In most situations, the effort required to revalidate the PHA will be significantly less than that required to conduct a new PHA. The alternative, to *Redo* the PHA from the beginning, is typically a more costly and time-consuming approach.

The degree to which the prior PHA can be used in the revalidation will depend upon a number of factors, including the quality of the prior PHA, the amount of change that has been implemented since the last PHA, and the adequacy with which the last PHA was documented. There are three courses of action that may be chosen, based upon these considerations. These courses of action are outlined below in order of increasing labor, resource, and time demands:

- *Update and Revalidate:* This is the expedient, incremental approach where the PHA need only be revised to reflect changes that have occurred and new learnings that have been gained since the prior PHA was conducted. The update effort may be relatively modest for some processes (especially established, mature processes subject to infrequent change). For a process involving no incidents or changes, it may only be necessary to affirm the continued validity of the prior PHA.
- *Retrofit, Update and Revalidate:* In this approach, an initial effort is needed to address one or more "repairable" defects in the prior PHA. Once the defects have been repaired, the PHA can be updated and revalidated as described above.
- *Redo:* This option is indicated when a repair of the PHA is impractical due to the nature, number, or magnitude of the defects in the prior PHA, the incidents or near misses that have occurred since the prior PHA, or the changes introduced since the prior PHA. In this situation, it may be more cost and resource effective to start from the beginning with a "blank sheet of paper."

Chapter 6 will discuss the above options in greater detail and provide implementation guidance.

2.4. Establishing the Revalidation Schedule

As Figure 2.1 illustrates, periodic PHA revalidations are an ongoing requirement for the life of the process.

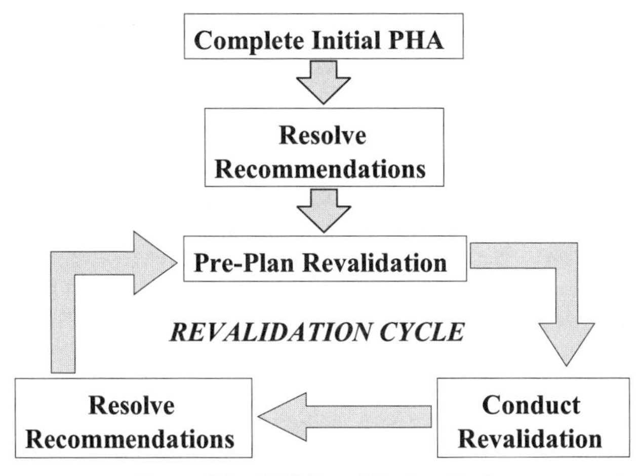

Figure 2.1. PHA Revalidation Cycle.

While many companies specify PHA revalidation every five years, specific facilities may elect to perform a revalidation sooner than this, for a number of reasons.[1]

Examples of situations where facilities may choose a more frequent revalidation include:

- Companies may decide that more frequent revalidations are more consistent with their loss prevention goals.
- If a major process or equipment revision is in progress, it may be more cost effective for a company to revalidate the unaffected portion of the process while performing the PHA of the modification;.
- MOC and prestartup safety reviews (PSSR) are intended to maintain the integrity of original safety features designed into the process, and to ensure that any new hazards are properly managed. However, the potential that overlooked and uncontrolled hazards exist, because of unidentified interactions, increases with the number of process modifications. Some companies may wish to consider triggering a revalidation based upon the cumulative number of changes.

[1] This book will commonly refer to a five-year revalidation cycle. This value, while specified in the OSHA and EPA regulations, is also the approximate median value recommended by industry associations, and represents a frequency that has been commonly applied by many companies in the past. Facilities not covered by the OSHA or EPA regulations should remember that they may establish their own, appropriate revalidation frequencies. As discussed here, regulated facilities may choose to revalidate more frequently than once every five years.

- Some companies have established frequencies for revalidating PHAs based upon a risk categorization (e.g., high, medium, low); for "high" risk processes, revalidation is sometimes more frequent than every 5 years. This type of approach is consistent with API Recommended Practice 750, *Management of Process Hazards* (API 1990).
- A significant incident or an unfavorable incident trend in a process might call into question the adequacy of the prior PHA, prompting an expedited revalidation. Similarly, incidents at another company site, or even outside the company, may foster reservations with regard to the prior PHA.
- Following a merger or acquisition, there may be a perceived need to reconcile quality or protocol disparities in PHAs.
- A company might have a concern that PHAs conducted early in the development of the facility PSM program may not have been as rigorously or as effectively conducted and may warrant review by more experienced PHA teams.

When Should the Clock Start?

What determines the required date for beginning the PHA revalidation meetings; is it 5 years from (1) the prior PHA first meeting date, (2) the prior PHA last meeting date, (3) the date the prior PHA report was issued, or (4) some other date? PHA meetings can span weeks and the final PHA report can be issued several weeks (if not months) later.

A prudent approach is to calculate the 5-year date from the first meeting date for the prior PHA. This approach assumes that the prior PHA team had up-to-date PSI at the start of their first meeting, but did not consider in their analysis changes and incidents that occurred during the conduct of the PHA (typically, the P&IDs and related PSI are "frozen" as of the start of the PHA team meetings). This approach helps ensure that changes and incidents that may not have been factored into the prior PHA are considered in the revalidation.

Regardless of the method used to calculate the 5 year revalidation date, the objective should be to ensure that the PHA revalidation team considers all changes and incidents in the process that were not considered in the previous PHA.

In summary, various factors may influence a company to perform PHA revalidations on a more frequent basis. This decision may be made on a process-specific basis to address needs unique to that process.

2.5. The Role of a Revalidation Procedure

Company or site procedures or standards often outline the steps to be followed in conducting PHAs. The benefits of such procedures include:

- Establishing schedules to ensure that PHAs are conducted in a timely fashion;
- Ensuring that the conduct and documentation of the PHA complies with pertinent company and regulatory requirements;
- Providing for a consistent content and format so that those using the PHA as an information source can anticipate what information will be available and know how to find it; and
- Establishing responsibilities for key roles with respect to the PHA element, including that for recommendation follow-up.

From the Workshop...

A corporate standard establishing procedures for conducting PHAs and PHA revalidations, including specific criteria for evaluating PHAs, was cited by Revalidation Workshop participants as an item believed likely to ease the task of future revalidations.

Many readers may find a company or site-specific revalidation procedure to be of value. While some organizations may need to address additional considerations, it is suggested that the revalidation procedure at least cover the procedural steps that are listed below and discussed in detail in this book:

- Preparing for the revalidation (Chapter 3);
- Evaluating the completeness and quality of the prior PHA (Chapter 4);
- Identifying changes and incidents that have occurred since the prior PHA was conducted (Chapter 5);
- Identifying an appropriate revalidation methodology (Chapter 6);
- Conducting the revalidation study sessions (Chapter 7); and
- Documenting the revalidation study (Chapter 8).

3

Preparing for the Revalidation Study

This book proposes one logical approach to the revalidation task, as depicted in Figure 3.1. This is not the only possible path to success; however, it is an approach distilled from numerous successful revalidation studies. This chapter discusses preparations important to the success of the revalidation effort.

3.1. Preplan the Revalidation

Proper preplanning of the revalidation is essential for a cost-, time-, and resource-effective effort. The preplanning may be accomplished by the study leader, with the support of other interested parties such as the process expert, production management, or the site PSM or PHA coordinator. Common issues to be considered when preplanning are discussed below.

3.1.1. Establishing the Scope of the Revalidation

It is imperative that the physical and analytical scope of the revalidation effort be clearly delineated so that affected equipment, personnel, records, etc. can be identified for the revalidation.

The scope of the revalidation need not be dictated by the scope of the prior PHA. For example, some organizations chose initially to divide a single process into more manageable subsections, with a PHA conducted for each subsection. It is conceivable that an organization may choose to otherwise aggregate (or perhaps further subdivide) processes prior to the revalidation effort. For example, it might be decided to group two PHAs (one nearly five years old and one only three years old) into a single revalidation PHA. There is no apparent impediment to doing so, as long as no single PHA goes beyond its revalidation deadline. In such circumstances, the revalidation team would need to gather information on both PHAs in order to plan the revalidation effort properly.[1]

It is not uncommon for specialized PHA considerations (e.g., siting or human factors) to be addressed via a separate study conducted in support of the main PHA.[2] It should be determined whether such adjunct studies are to be included in the revalidation effort, or whether they will be separately revalidated.

3.1.2. Selection of Team Members

There are two schools of thought with respect to selection of PHA revalidation team members; either, (1) for efficiency, keep the same team as before, or (2) for objectivity, select a new team. Either approach can be defended on its merit. In reality, however, organizational issues (e.g., loss or reassignment of past team members) often prevail, and the team ends up as a mix of new personnel and personnel that were involved in the prior PHA. Furthermore, experience with the prior PHA may indicate the need to supplement the revalidation team with members representing specific areas of expertise not present on the prior team (e.g., metallurgist, control engineer, R&D chemist, loader/unloader).

From the Workshop...

Revalidation Workshop participants noted that a mix of both new and prior team members might satisfy at least three goals:

- Providing some consistency of approach to the PHA effort;
- Providing fresh insight and objectivity; and
- Training participants for future PHA revalidations

As is the case with any initial PHA, the responsibility rests with the study leader to ensure that a proper mix of knowledge, experience, and training is embodied in the team make-up.

Those facilities covered by the OSHA or EPA regulations should keep in mind the specific requirements for team make-up that are discussed subsequently in Section 4.1.3.

[1] Situations where multiple PHAs exist for a single process could include projects for which PHAs were staged as different portions of the design were completed, or where a separate PHA was conducted for a control system upgrade. The revalidation PHA offers an opportunity to pull these multiple PHAs together.

[2] Those processes covered by the EPA RMP Rule are subject to stationary source siting requirements that are broader than the facility siting evaluation requirements established by the OSHA PSM Standard. (See Appendix A.) Where the terms "siting" is used in this book, it should be interpreted as encompassing either or both of the OSHA facility siting and the EPA stationary source siting requirements, as appropriate.

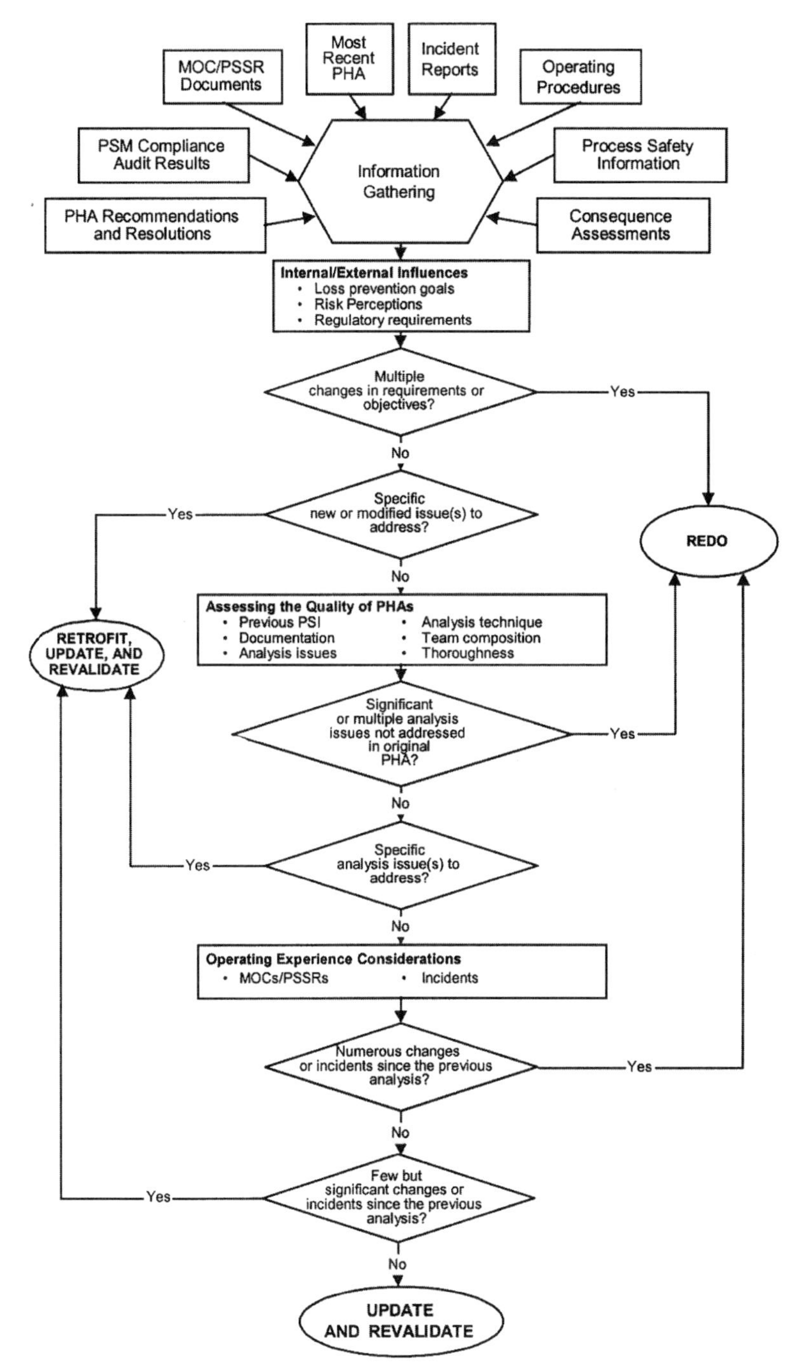

Figure 3.1. PHA Revalidation Logic Flowchart.
(Adapted from Crumpler and Whittle, 1996.)

3.1.3. Scheduling—Estimating Time and Resources

An important part of the preplanning is to establish a schedule that ensures the completion of the revalidation to meet any company or regulatory deadlines. The time needed to complete a revalidation will depend upon factors such as:

- The quality and completeness of the prior PHA;
- Availability of information from the prior PHA;
- The number of undocumented changes, if any, to be reviewed; and
- The quality of the MOC and PSSR programs.

PHA quality and completeness is likely to be one of the more important factors on the list. Refer to Chapter 4 for a discussion of factors that could negatively impact upon the prior PHA.

Experience to date allows us to place only broad boundaries on estimates of the time or resource requirements for completing a PHA revalidation. Significant content or quality problems will necessitate a *Redo* for some PHAs; for others, the number of changes to be evaluated will make a *Redo* the more cost-effective option. The time involved in completing a *Redo* should be comparable to that for an initial PHA.

For those situations where the *Update and Revalidate* option or the *Retrofit, Update and Revalidate* option are feasible, the time required for the revalidation sessions can range from 10% to 50% of the time needed for an initial PHA. Including the data gathering and preparation time, the total person-hours required for the revalidation can range from 35% to 75% of that required for an initial PHA.

Clearly, there is no such thing as a "typical situation"; however, the above figures can be used for a first approximation of the time requirements.

Another approach to estimating meeting time considers the number and complexity of the changes to be evaluated during the revalidation. For example, assume the changes are grouped into the following four categories for estimating evaluation time: small (5 minutes or less); medium (15 minutes); large (30 to 60 minutes); and very large (greater than 60 minutes). When grouping changes, it is advisable to review in more detail any change that will take longer than 60 minutes to get a more accurate time estimate. Table 3.1 gives typical examples of changes that may fit into each of these categories (Smith and Whittle, 2000).

These time estimates assume that all the modifications to the process (and any associated documentation) have been reviewed prior to the revalidation meetings by the PHA revalidation leader or other experienced personnel. This ensures that personnel who are very familiar with the modifications can explain to the full revalidation team the rationale for each change.

TABLE 3.1
Example Modification Categories for Estimating PHA Revalidation Meeting Time

Category	Modification
Small (5 minutes or less)	• Replace a carbon steel coupling with a stainless steel coupling • Change the sampling frequency on a product tank from once per shift to once per day • Install an additional block valve on the drain line from a heat exchanger
Medium (15 minutes)	• Replace an existing gas-fired reboiler with a shell/tube heat exchanger • Install a new fixed firewater nozzle on the west side of the process • Replace a rupture disk with a relief valve on a vessel • Install a 2-inch temporary hydrogen line from the hydrogen header to a reactor
Large (30 to 60 minutes)	• Install a new storage tank in the tank farm • Tie-in an atmospheric vent line from a reactor to an existing line to a thermal oxidizer • Remove a nitrogen regulator from a unit's nitrogen supply header • Install a new bypass line around the dehydration and scrubbing system from a reactor
Very Large (> 60 minutes)	• Install an additional process train that includes a line with a pump and two columns • Add a new scrubber tower and associated caustic circulation system

If the study worksheets (e.g., HAZOP worksheets) for the prior PHA are to be updated, the PHA revalidation leader may flag in the worksheets appropriate deviations or questions that may need updating based on this preliminary assessment of each modification. This preparation work will help accelerate the pace of the PHA revalidation meetings and allow the PHA revalidation leader to methodically guide the team through this task.

3.2. Identify, Collect, and Prepare Needed Information

Considerable information may be required to support a rigorous PHA effort. It should not be surprising that similar information might also be required for the PHA revalidation. In fact, the revalidation effort will likely require information not available during the prior PHA, such as the prior PHA itself, along with supporting information such as records documenting recommendation resolution.

3.2.1. Determining Information Requirements

A key step in preparing for a revalidation is to identify, collect, and prepare the information needed to support the analysis. Table 3.2 suggests types of information that would typically be considered. This listing has been compiled from a number of sources. All information types may not be applicable to a particular organization or revalidation, and some information may not be available.

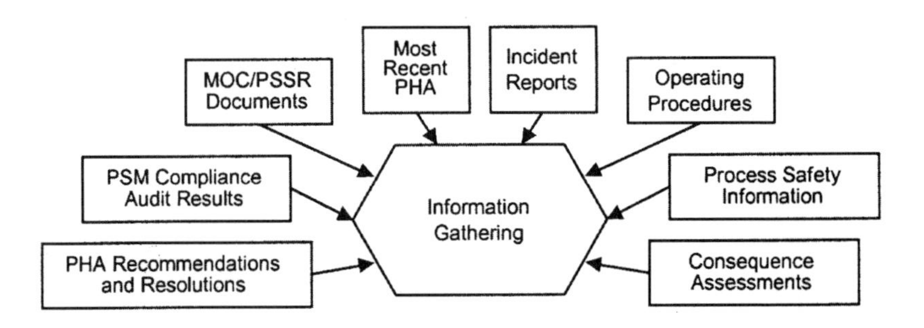

It is the responsibility of the study leader (or a designee) to ensure that the current process safety information (PSI) is adequate for revalidating the PHA. Incorrect or missing PSI could result in the PHA team not identifying existing hazards or, conversely, taking credit for safeguards that are not effective (e.g., engineering or administrative controls that have been removed or disabled). Although experienced personnel on a PHA team may compensate for some PSI shortcomings, it is important that such information shortcomings be identified and corrected to both support an effective revalidation effort and to comply with applicable regulatory requirements.

For maximum efficiency, PSI deficiencies should be resolved ahead of the revalidation meetings whenever possible. However, in some situations, missing or ambiguous information may have to be redeveloped during the revalidation.

3.2.2. Distribution of Information

Certain key information items can be provided to the team members for their review prior to the revalidation sessions. This could help team members prepare ahead of time so that they are more effective. Such information

TABLE 3.2
Documentation for Review during Preplanning a PHA Revalidation

Information Type	Comments
1. Prior PHA report(s) and related documentation	Establishes part of the base-line for revalidation. May be more than one report involved. Include reports for projects and changes (See MOC/PSSR).
2. Resolution completion report(s) for prior PHA recommendations (for all reports gathered above).	Establishes part of the base-line for revalidation. Recommendations remaining open should be resolved or factored into the revalidation.
3. Results from informal safety reviews	May reveal system weaknesses not detected via PHA or incident investigations.
4. Safety committee reports	May be a source of operational concerns.
5. MOC and PSSR documentation • For process and equipment to be studied • For interconnected processes and equipment, as appropriate	Identifies the controlled changes made since last revalidation. Database against which uncontrolled changes, if any, can be identified.
6. Process safety management (PSM) system audit results	May identify weakness in PHA, PSI, MOC, or PSSR programs, etc. May identify specific weaknesses in the prior PHA itself.
7. Audit results by outside agencies	See 6.
8. Accident and near miss reports • For process to be studied • For similar processes at other sites	May identify weaknesses in PSM program, or prior PHA. Source of additional scenarios to consider during revalidation. Should cover last five years, or period since last revalidation.
9. Process upset report and investigations	Provides information on events that didn't trigger the near-miss criteria, but may still be indicative of a potential process problem.
10. Piping and instrument diagrams • Current, up-to-date • Those used for prior PHA, if available	Serves as "road map" for study sessions. Comparison of current P&IDs with those of prior PHA helps identify potential changes.
11. Node descriptions for prior PHA	Provides information on the conduct of the prior PHA.
12. Block flow or process flow diagrams	Describes the process or facilities under study.

Continued on the next page

Table 3.2 (Continued)

Information Type	Comments
13. Area electrical classification drawings	Describes the process or facilities under study.
14. Process chemistry description	Describes the process or facilities under study.
15. Safe operating limits for the process	Describes the process or facilities under study.
16. Material and energy balances • Current, up-to-date • Those used for prior PHA, if available	Describes the process or facilities under study.
17. Descriptions of safety systems and records of their functional testing	Safety systems are commonly relied upon safeguards. Documentation is required to ensure that claimed safeguards are actually in place and can be relied upon.
18. Operating procedures • Current, up-to-date • Those used for prior PHA (if available)	Valuable process reference. Can be used to identify nonroutine operating modes for analysis, as well as any uncontrolled or undocumented changes.
19. PSV documentation, sizing calculations and test reports	Provides information on critical safeguards
20. Mechanical integrity information • Maintenance records and procedures • Equipment inspection reports	May assist in evaluating equipment reliability, especially for critical safety systems.
21. Safety and emergency maintenance work orders	May help identify undocumented changes.
22. Consequence assessments • Siting studies • EPA RMP WCS and ARS results	Can "calibrate" the team with respect to how serious the effects of a release could be; assists in consequence/risk ranking.
23. Material safety data sheets	Property data on chemicals used in the process, service systems, etc.
24. Other pertinent process safety information as defined in 1910.119(d)	May include plot plan drawings, critical instrument lists, etc. Can aid in identifying any new or previously overlooked hazards. May identify uncontrolled or undocumented changes.

might include copies of the prior PHA and pertinent P&IDs. The value of distributing such information can be evaluated on a case-by-case basis.

3.3. Review and Analyze Information

The review of the PSI will help define the scope and magnitude of the revalidation task. While primary focus will likely be placed on the P&IDs, other information such as maximum intended inventories, safe operating limits, relief system design and design basis, should not be overlooked. The study leader (or a designee) may compare the relevant PSI used in the prior PHA and the current PSI to identify differences that may reflect changes to be reviewed during the revalidation. Detailed procedures for identifying changes will be discussed in Chapter 5.

Note that, for certain types of information, it is important to gain a perspective of not just what information is currently available but, additionally, of *what information was available when the prior PHA was conducted*. To borrow a phrase from the computer industry... "Garbage in, garbage out." A PHA that was based on inadequate, insufficient, or incorrect information (e.g., outdated or incomplete P&IDs) is not likely to be suitable to serve as the firm foundation for a revalidation effort. Consequently, any information deficiencies that apparently existed at the time of the prior PHA should be noted. *Redoing* a PHA may be the most appropriate option if there are compelling indications that significant hazards of the process may have been overlooked, underestimated, or not properly controlled as a consequence of inaccurate or incomplete PSI being used during the prior PHA.

A number of types of information having particular relevance to the revalidation effort are addressed below, beginning with the prior PHA itself.

3.3.1. Prior PHA Report(s) and Related Documentation

The prior PHA serves as the starting point of the revalidation effort. In order to scope out the revalidation effort, it is first necessary to evaluate the quality of the prior PHA. Are there gaps and deficiencies in the prior PHA? In some cases, it will be determined that the prior PHA met all company and regulatory requirements, was generally well conducted, and accurately assessed the hazards of the process as they were understood at the time. If so, as it stands, this PHA would serve as a suitable baseline for the revalidation effort. In other situations, some *Upgrade* (or *Retrofit*) of the prior PHA may be needed during the revalidation effort. More specific guidance on evaluating PHAs is provided in Chapter 4.

Reference has been made to the "prior PHA report" and, in many circumstances, there will be only one such document to consider. However, as discussed in Section 3.1.1, there may be circumstances where two or more prior PHAs are aggregated for revalidation. A more commonly encountered circumstance might be where a series of "mini-PHAs" have been conducted as part of the evaluation of changes under the MOC program. Alternatively, a "full-blown" PHA may have been conducted for a major project that modified the process or equipment.

Whatever the nature of the ancillary PHAs, it is important that the scope of the revalidation be clearly defined (i.e., what processes and equipment are covered) and that all relevant PHAs falling under that scope be gathered and factored into the revalidation effort.

3.3.2. Resolution Completion Report for Prior PHA Recommendations

The prior PHA(s) likely proposed a number of recommendations intended to mitigate the hazards identified and evaluated during the study. The revalidation team should review and evaluate the status of these recommendations. It is important to understand the changes that have been implemented since the prior PHA. If the prior PHA recommended a new hazard control (e.g., a new shutdown system), was the system actually installed? Can credit be taken for the intended new protection when revalidating the PHA? Did the manner in which the recommendation was implemented actually provide the intended protection; i.e., was it effective in addressing the original concern identified in the prior PHA?

It may be appropriate to verify the actual implementation of the recommendation using a document review and/or field inspection. If the status of a recommendation cannot be verified prior to the study sessions, the revalidation team should review the recommendation during the revalidation.

Many companies are using the PHA revalidation process to reassess the need for any unresolved recommendations remaining from the prior PHA. Some of these recommendations may not have been related to process safety. For example, process improvement (operability) recommendations, and recommendations addressing conventional worker safety issues (as contrasted with process safety issues) are often lumped together with the process safety recommendations in the PHA report. The company may choose the PHA revalidation as an opportunity to document that these recommendations are being redirected to other tracking systems separate from the PHA system. Conversely, the team should also be alert to recommendations erroneously classified as either operability issues or conventional worker safety issues, but actually having process safety significance. These should be appropriately reclassified for follow-up.

The review of other recommendations may determine that they are no longer required. The revalidation team may recommend closing out these recommendations, and documenting the rationale for not implementing them. Considerable thought likely went into these recommendations and the revalidation team should not arbitrarily discard them without first attempting to understand their basis. Nonetheless, recommendations still open five years after the prior PHA should be reevaluated to determine their true criticality.

There may be relevant process safety-related recommendations remaining that, for one reason or another, have not been implemented. These recommendations, if they cannot be implemented prior to the revalidation, should be rolled into the PHA revalidation deliberations.

Finally, the PHA revalidation provides an opportunity to evaluate the manner in which recommendations were resolved to ensure that the PHA revalidation team agrees with the actions taken to implement the recommendations or the rationale for not implementing them.

3.3.3. MOC and PSSR Documentation

Process units may be modified over time for a number of reasons (e.g., technology improvements, capacity increases). Properly implemented, the MOC and PSSR elements can be the "glue that holds things together" between revalidations. In the Preamble to the PSM Standard, OSHA justified the 5-year revalidation frequency based upon the belief that MOC and PSSR would control change and guard against major degradation of process safety posture pending the next revalidation.

As noted in Chapter 2, when many changes occur over time there is a potential for synergistic effects that may not be detected under the MOC system. Thus, it is

From the Workshop...

Greater emphasis on the quality of PSM program implementation (especially the PHA, MOC, and PSSR elements) was perceived to be key to facilitating future PHA revalidations.

Revalidation Workshop participants identified MOC system improvements in particular as one of the more significant activities that could ease the revalidation task. Specific suggestions included:

- Ensure that the MOC form provides for a good, concise, single line description of the nature of the change (to aid in screening MOCs);
- Provide a checklist on the MOC form to ensure that all PHA "essential criteria" are addressed during the evaluation of the MOC (see section 4.1, below); and
- Place more emphasis on MOC (and PSSR) system training to ensure their effective implementation.

prudent to look at these changes globally during the revalidation effort. Did one MOC violate the assumptions of another? Were there any new hazards or interactions caused by the changes when taken as a whole? The likelihood that such hazards exist increases with the number of process modifications.

As part of the study preparation, the study leader (or designee) should accumulate relevant MOC and PSSR documentation pertaining to the process under study. This documentation will describe the *controlled* changes made since the last revalidation. Once this database is assembled, it serves as a reference against which *uncontrolled* changes, if any, can be identified (this will be discussed further in Chapter 5).

Outstanding recommendations from the MOCs or PSSRs should be reviewed in a manner analogous to that previously described for PHA recommendations. Additionally, the revalidation team should note for review documented changes that do not appear to have been adequately reviewed or that otherwise do not meet the requirements of the facility MOC procedure.

Table 3.2 also lists MOC and PSSR information for other processes and equipment interconnected to the process under study. It should be remembered that changes in service systems (e.g., plant chilled water system) or other processes might have an impact on the process under study. Such relevant changes need to be identified and considered during the revalidation.

3.3.4. PSM System Audit Results

PSM system audit results may point to weaknesses in the PHA program, or the prior PHA itself, that might not otherwise have been detected. Similarly, the audit may identify problems with the MOC or PSSR systems relied upon to control change between revalidations.

Some facilities may have received other, external audits; e.g., under state-level chemical accident prevention programs, or under EPA's RMP Rule. If so, such audit reports may also be considered during the information-gathering phase of the revalidation task.

It should be noted that, typically, PSM system audits are conducted using some sort of sampling scheme to identify documentation that is to be reviewed. Thus, the fact that a particular PHA report is not mentioned in an audit finding should not be construed as conclusive proof of the quality of that particular PHA; the PHA may have not been examined as part of the audit.

3.3.5. Incident and Near-Miss Reports

Incident and near-miss reports may identify event scenarios that warrant consideration within the revalidation effort and should be reviewed as part of the preparation. Where relevant information can be gathered on incidents in similar facilities or processes, such information may also be considered.

A major incident, a series of less significant incidents, or numerous near-misses in a process unit can be indicators of a weakness in, or failure of, some PSM system element(s). This need not necessarily be a problem with the PHA element, but incidents involving the subject process should be scrutinized to see if the particular circumstances prompt a concern with the quality of the prior PHA. For example, the PHA team may not have identified the potential for a particular incident; alternatively, the PHA team may have identified the potential for the incident, but erroneously judged the safeguards to be adequate.

From the Workshop...

As noted in Table 3.2, some companies track abnormal occurrences in the process that may not exceed their criteria for investigation as near-misses. Significant trends of such occurrences may be indicative of potential process safety issues warranting consideration during the revalidation. Workshop participants suggested that process or batch records could be reviewed to track actual process performance versus anticipated performance in order to identify such abnormal occurrences.

The recommended corrective actions documented in the incident investigation reports, and the details of their resolution, should be reviewed in a manner analogous to that previously described for PHA recommendations.

3.3.6. Piping and Instrument Diagrams (P&IDs)

P&IDs are typically the focal point of the PHA meetings, and performing a revalidation based upon incorrect or out-of-date drawings would be a waste of time and resources. Particular care should be devoted to ensure that the P&IDs clearly and accurately reflect the process and equipment details needed to support the revalidation effort. While the cost of field verification and update of these drawings can be significant, if the condition of the P&IDs requires such updates, it is imperative that this be accomplished prior to the start of the revalidation.

Many organizations include in the PHA documentation copies of the P&IDs used, marked to identify the individual study sections. As will be discussed in Chapter 5, comparison of the current and prior P&ID revisions can be used to identify changes that have occurred since the prior PHA was conducted.

3.3.7. Operating Procedures

It is not uncommon for initial PHAs of continuous processes to focus only on normal operations, failing to address nonroutine, critical operating modes such as startup, shutdown, preparation for maintenance, emergency operations, emergency shutdown, and other activities whose characteristics may differ considerably from normal operations. Similarly, PHAs of batch processes may neglect operations such as vessel clean-outs, water runs, or pressure testing.[3]

Experience indicates that many accidents do not occur during "normal" operation but, rather, during such nonroutine modes of operation. Consequently, it is important that a PHA evaluate the hazards of a process during nonroutine as well as normal (routine) operating modes. Careful review of the procedures for all modes of operation can help to identify these nonroutine operations. Additionally, discussions with revalidation team members may identify other such nonroutine operations.

[3] CCPS *Guidelines for Process Safety in Batch Reaction Systems* provides guidance on the process safety considerations unique to batch processes.

4

Evaluating the Prior PHA Study

The primary objective of a PHA revalidation is to produce an updated PHA that adequately identifies, evaluates, and controls the hazards of the process, *as they are currently understood.* Preferably, this effort will be an incremental effort; that is, the prior PHA will only need to be updated to reflect changes that have occurred since the PHA was initially conducted.

The presumption underlying the incremental approach is that the prior PHA *adequately addressed the hazards of the process at the time the PHA was completed.* Only if this is true, is the incremental approach valid. Therefore, a key step in planning a PHA revalidation is to evaluate the adequacy of the prior PHA to determine if it is a suitable baseline upon which to build the revalidation effort.

From the Workshop...

Revalidation Workshop participants noted that the evaluation of the prior PHA was an opportunity to perform not just a quality assurance check on that particular PHA but, potentially, the PHA program as well. Integration of the learnings from multiple revalidations may point to opportunities to improve the implementation of the facility PHA effort.

There are two broad considerations to address in evaluating the prior PHA. The first deals directly with the quality of the work product; that is, did the prior PHA adequately evaluate the hazards of the process? For example: Was the review sufficiently detailed? Were correct conclusions drawn? Was the PHA technique correctly used? A higher quality PHA will need less remedial effort during the revalidation.

The second consideration applies where applicable specifications or standards exist that detail criteria for the conduct of PHAs. In such situations, it is necessary to ask: Did the prior PHA meet the applicable criteria?

This chapter describes an approach to evaluating the prior PHA with respect to both of these considerations. While this chapter is written as if a single PHA were involved, the considerations outlined here can be applied to as many PHAs as there are to be included in the revalidation.

4.1. Evaluation of the PHA with Respect to Essential Criteria

Of the two considerations, compliance with established criteria is the more basic, and will be addressed first. To do so, it is necessary to answer three questions:

1. What are the criteria?
2. Did the PHA meet the criteria?
3. If not, what needs to be done to address the deviations from the criteria?

The answer to the third question helps define the scope of the revalidation task. It is necessary to distinguish between those deviations (or problems) that can, and those that cannot, be repaired in the prior PHA. Repairable problems can be addressed within the revalidation effort without having to *Redo* the PHA. Nonrepairable problems represent issues with the conduct of the prior PHA that are so basic that they cannot be addressed without *Redoing* the PHA.

For example, suppose that the company's internal PHA policy document requires that details of PHA recommendation resolution be documented as an addendum to the original PHA report, but that this has not been done. The PHA report could be "repaired" by creating this addendum and adding it to the report. In contrast, suppose that the company's policy requires that all PHAs be conducted using the HAZOP technique. A PHA conducted using the What If technique likely could not be "repaired" in a manner that transforms it into a HAZOP, without actually conducting a HAZOP analysis (i.e., without *Redoing* the analysis).

Many companies have established internal criteria for the conduct of PHAs. Facilities covered by regulatory requirements are further subject to criteria established by the regulations. For purposes of illustration, the discussion below will describe those criteria established for the conduct of PHAs under the OSHA PSM Standard and the EPA RMP Rule.[1] There are four criteria, each tied directly to an explicit regulatory requirement. Three of these criteria deal primarily with the way that the PHA was conducted, and the fourth deals with the PHA documentation.

[1] Facilities not covered by the OSHA PSM or EPA RMP regulations can voluntarily use these requirements as a basis for their PHA program if they so choose.

4.1.1. PHA Rigor

The OSHA PSM and EPA RMP regulations require that:

> The process hazard analysis shall be appropriate to the complexity of the process and shall identify, evaluate, and control the hazards involved in the process.

This requires a subjective determination: Was the PHA sufficiently rigorous or was there a clear mismatch between the level of review given and the complexity and the degree of hazards present in the process? No quantitative criteria can be proposed for this decision, but some indicators can be suggested:

- How complex is the process with respect to chemistry, controls, safety systems, or sequence of operations? How severe are the hazards inherent in the process; i.e., how bad could a bad incident be? How close does the process operate to its safe operating limits? How sophisticated are the controls and safeguards provided for the process? Processes that are complex, highly hazardous, or that operate "close to the edge" would normally dictate the need for a more rigorous PHA technique. The use of a simple technique (e.g., checklist) or the cursory application of a more detailed technique (e.g., HAZOP) during the prior PHA may call into question the adequacy of the prior review.
- How many incidents and near-misses have occurred since the prior PHA was conducted? Were these scenarios that were demonstrated "in real life" identified in the prior PHA? PHAs are intended to identify and provide controls for those scenarios that could lead to incidents and near-misses. While it is recognized that no PHA team is likely to "find everything," an unsatisfactory history of incidents or near-misses could be indicative of an inadequate PHA (e.g., an inappropriate technique was used or the combined experience of the team was not sufficient to discover the potential incidents). Alternatively, the incident(s) could have resulted from the inadequate implementation of the PHA learnings.

A significant mismatch between the depth and rigor of the PHA and the complexity of the process may constitute a nonrepairable problem, requiring that the PHA be redone.

Guidance on PHA technique selection, with respect to the nature of the process and the perceived magnitude of the hazards, can be found in the *HEP Guidelines*.

4.1.2. Methodology Used

The regulations also require that:

> The employer shall use one or more of the following methodologies that are appropriate to determine and evaluate the hazards of the process being analyzed.
>
> (i) What-If;
> (ii) Checklist;
> (iii) What-If/Checklist;
> (iv) Hazard and Operability Study (HAZOP);
> (v) Failure Mode and Effects Analysis (FMEA);
> (vi) Fault Tree Analysis; or
> (vii) An appropriate equivalent methodology.

If a technique not listed above was used for the prior PHA, this may constitute a nonrepairable problem requiring that the PHA be redone.[2]

4.1.3. Team Make-Up

The PSM and RMP regulations further require that:

> The process hazard analysis shall be performed by a team with expertise in engineering and process operations, and the team shall include at least one employee who has experience and knowledge specific to the process being evaluated. Also, one member of the team must be knowledgeable in the specific process hazard analysis methodology being used.

It is unlikely that failure to include the right mix of experience and expertise on the team could be corrected post-analysis without redoing the PHA.

4.1.4. Documentation

The regulations require that:

> Employers shall retain process hazards analyses and updates or re-validations for each process covered by this section, as well as the documented resolution of recommendations ... for the life of the process.

Problems with documentation can possibly be repaired, if other sources of information allow the missing documentation to be reconstructed or sup-

[2] OSHA and EPA do allow the use of an *"appropriate equivalent methodology"* but have provided little information on what such a methodology might be. One might reason that the other techniques described in the CCPS *HEP Guidelines* would be suitable for consideration; however; the chosen *"appropriate equivalent methodology"* may have to be defended.

plemented. However, substantive voids in the documentation that cannot be filled leave in question the adequacy of the prior PHA and it may be necessary to *Redo* the PHA to fill such voids. See Chapter 8 for additional guidance on documentation issues.

4.1.5. Drawing the Conclusions

In summary, company or external (e.g., regulatory) criteria may exist dictating the manner in which the PHA is to be conducted. Where binding criteria exist, it is necessary to evaluate the prior PHA against these criteria to determine whether it can suitably serve as the basis of the revalidation effort. Some deviations from the established criteria can be corrected (repaired) to permit the use of the prior PHA; more substantive problems likely cannot be repaired and the PHA would have to be *Redone*.

Some organizations find it useful to provide a checklist for revalidation teams to use in evaluating the prior PHA. An example of such a checklist, adapted from a checklist used by one CCPS member company, is provided as the *Essential Criteria Checklist* in Appendix B. This particular checklist was developed specific to the company and regulatory criteria applicable to the user; readers may wish to develop similar checklists specific to their own company or facility needs and requirements.

A PHA may comply with established requirements such as those discussed above, and yet be a poor candidate for serving as the basis of a revalidation effort. It is next necessary to look at the quality of the prior PHA.

4.2. Evaluation of PHA Quality and Completeness

Did the prior PHA adequately address the hazards of the process? Some organizations, in an effort to ensure an affirmative answer to this question, have created criteria against which a PHA may be measured. Additional criteria may also be established by regulatory agencies; for covered facilities, these criteria include those established by the OSHA PSM Standard or the EPA RMP Rule.

"Once we know the essentials are there, it is essential that we look at the quality"

This section discusses the use of screening criteria for quality and completeness that could be applied to the baseline PHA to assess how well, and how completely, it addressed the hazards of the process. A reasonable number of unsatisfied criteria can be addressed during a *Retrofit, Update and Revalidate* of the PHA. However, if a large number of quality and completeness criteria are not satisfied, careful consideration should be given to *Redoing* the PHA. Similarly, the effort needed to *Retrofit* a single criteria might be so great as to suggest that a *Redo* might be the more effective use of resources.

For purposes of illustration, the discussion below will describe criteria established under the OSHA PSM Standard for the conduct of PHAs (similar criteria are contained in the EPA RMP Rule, see Appendix A). The PSM Standard requires that:

> The process hazard analysis shall address:
>
> (i) The hazards of the process;
> (ii) The identification of any previous incident which had a likely potential for catastrophic consequences in the workplace;
> (iii) Engineering and administrative controls applicable to the hazards and their interrelationships such as appropriate application of detection methodologies to provide early warning of releases...;
> (iv) Consequences of failure of engineering and administrative controls;
> (v) Facility siting;
> (vi) Human factors; and
> (vii) A qualitative evaluation of a range of the possible safety and health effects of failure of controls on employees in the workplace.

Potential inadequacies in the prior PHA are classified as either gaps or deficiencies. Gaps occur when company, or regulatory, requirements are substantially (or, in the extreme, totally) omitted from the prior PHA. Deficiencies can be errors in applying the selected PHA methodology, or they can involve the incomplete consideration or treatment of a requirement during the PHA study.

A failure to consider past incidents, for example, is a gap that could readily be addressed and corrected by the revalidation team when *Retrofitting* the PHA. Similarly, as noted in Appendix A, the EPA RMP Rule has broader requirements than those contained in the PSM Standard; e.g., the requirement to address the "*qualitative evaluation of a range of the possible safety and health effects*" includes off-site health effects, not just those pertaining to employees. Those processes covered by the RMP Rule may need to fill this gap by *Retrofitting* their PHAs to extend the consideration of safety and health effects beyond the fence line.

The revalidation team might address other gaps on a stand-alone basis. For example, it is not uncommon for a facility PHA to be supplemented by a separate, stand-alone, siting review.

Examples of deficiencies include not consistently considering the failure of engineering and administrative safeguards when developing consequences and not consistently applying the selected PHA method. Correction of deficiencies requires *Retrofitting* the prior PHA study. If the number or magnitude of the deficiencies is severe, it may be more expeditious to *Redo* the study.

Some companies have developed a checklist for revalidation teams to use when screening the prior PHA for gaps and deficiencies. Such a checklist could reflect both company-specific criteria, as well as any applicable regulatory criteria. An example of such a checklist is provided as the *PHA Quality and Completeness Checklist* in Appendix C. Readers may wish to develop similar checklists specific to their own company or facility needs and requirements.

4.3. Other Considerations

It is not uncommon for project-related PHAs to be performed on a process that has not yet been operated. In such situations, at least some of the PHA deliberations were likely to have been somewhat conjectural. It is conceivable that, under such circumstances, the PHA team could have either inadvertently overlooked hazards (because the projected unit response was based on design concepts and not on actual operation), or postulated hazards that did not materialize. In either case, the results of the initial PHA are likely to need review and revision in order to represent accurately the current knowledge regarding the process hazards and the engineering and administrative controls that are in place. *Redoing* the PHA may be the most appropriate choice when current operating practices or conditions significantly diverge from those addressed by the prior PHA.

From the Workshop...

Revalidation Workshop participants suggested that valuable insight into the conduct of the prior PHA might be gained through discussions with the prior team leader or members. Additionally, it was suggested that comparison with other PHAs conducted on similar processes might aid in evaluating the quality and completeness of the prior PHA.

4.4. Common Problems with PHAs

It may be helpful for those responsible for screening the PHA quality to be aware of some of the problems that are more commonly encountered with PHAs. If a root cause analysis of such shortcomings were to be performed, the results would likely reveal that the problems result from factors such as:

- An inexperienced team leader;
- A PHA team that was not appropriately staffed or properly trained for the review;
- Poor preparation prior to the PHA review; or
- Failure to allow sufficient time for the review resulting in a cursory treatment.

PHA revalidations are equally subject to quality problems resulting from these same, and similar, causes. It is a responsibility of the team leader to ensure that such problems do not arise in the revalidation. PHA revalidations are intended, in part, to address past PHA deficiencies, not to create new ones.

While the following is not purported to be a comprehensive list, it is indicative of some of the problems that might be encountered with the prior PHA:

- Inadequate team and/or facilitator expertise;
- Inadequate Process Safety Information (e.g., incomplete or inaccurate P&IDs);
- Failure to identify or document all credible hazards associated with the process;
- Important initiating events not captured;
- Failure to address all operating modes (e.g., start-up, shut-down, clean outs, catalyst changes);
- Missing node descriptions;
- Consequences not carried to their "global" conclusion (e.g., stopping at "high pressure in vessel" rather than considering the possibility of "vessel rupture" and its subsequent consequences);
- Taking credit for existing safeguards when evaluating consequences;
- Claiming ineffective safeguards when evaluating likelihood (e.g., taking credit for operator intervention when, in reality, the event would develop too rapidly for the operator to respond);
- Failure to specifically identify engineering or administrative controls (e.g., by instrument ID number or procedure number);
- Claiming invalid or nonsensical safeguards (e.g., "that has never happened here");
- "Rubber stamping" safeguards (e.g., overuse of "Operator Training and Knowledge");

- Failure to detect common mode failure issues between control and shutdown systems;
- Failure to address siting or human factors;
- Overly optimistic estimation of consequences or likelihood, and corresponding underestimation of risk;
- Inconsistent risk ranking of scenarios;
- Failure to consistently consider past incidents;
- Documenting the study by exception (i.e., only documenting those deviations or scenarios for which severe consequences are likely... reviewer cannot tell whether other scenarios were overlooked, or considered and discounted);
- Formulation of recommendations that are either too vague or otherwise impossible to implement;
- An improbably small number of recommendations indicating, perhaps, a too-cursory analysis; and
- Inadequate documentation of the review (e.g., no explanation of the methodology used by the prior PHA team).

5

Identifying Changes That Have Occurred Since the Prior PHA

As previously noted one of the basic reasons for PHA revalidation is to address changes that have occurred since the prior PHA was conducted. Consequently, one of the most important, and time-consuming, preparatory steps for the revalidation effort is the identification of these various changes. A full understanding of the number and nature of these changes is key to determining the scope of the revalidation effort, and to completing an effective revalidation.

Ideally, all changes would be carefully controlled and implemented via the MOC and PSSR systems; however, some changes may have slipped through the system. It is important that changes, whether controlled or not, and whether documented or not, be identified for consideration during the revalidation effort. This chapter will address the task of identifying both controlled, and uncontrolled, changes.

Table 5.1 lists examples of the types of changes that may be encountered. Experience has shown that there are a number of particularly productive sources of this change information. These will be discussed in greater detail below along with suggested approaches for identifying the changes. While these recommended approaches should apply to many situations, it is recognized that documentation practices vary from one company to the next, and not all approaches will apply in all situations.

5.1. Logging the identified changes

Identifying process changes and gathering and organizing relevant information about them for use in the revalidation is an important part of preparing for a PHA revalidation. Appendix D contains an example *Change Summary Worksheet* that could be used as an aid for this task. This worksheet is organized by P&ID number (first column) so that the Revalidation Team can evaluate each change in the context of the P&ID-based review during the revalidation work sessions. The worksheet provides for documentation of the nature of the change, the source of information regarding the change (i.e., in what documentation was the change discovered), and explanatory comments. Finally, the last column allows documentation of the proposed action for addressing the change during the revalidation.

There often may be no action required to address a particular change during the revalidation. However, those changes that did not receive adequate review prior to implementation (e.g., did not go through the MOC process or only received a cursory review) should be flagged for more careful consideration during the revalidation.

5.2. Documented and Controlled Changes

The simpler task, that of identifying documented changes, will be addressed first. A number of common sources of information should be considered, as described below. Guidance will be provided in the next section regarding the search for *undocumented* changes.

5.2.1. MOC and PSSR Review

By their very nature, MOC records should be a significant source of information about changes to the process and equipment. A search of the MOC files by process or equipment identification number should readily identify these *documented* changes.

Just compiling copies of the relevant MOC records is not enough. Having an MOC form does not guarantee that the described change was adequately reviewed. The content of the forms should be reviewed to confirm that adequate consideration was given to the nature and potential safety impact of the change.

There is typically a correspondence between MOCs and the PSSRs that should follow them; although, not all PSSRs are generated in conjunction with an MOC form. A review of PSSR records can serve as one means of cross-checking the MOC records and identifying changes that did not receive proper MOC review.

Special attention may be warranted for those processes where the PHA was performed early in the development of the facility PSM program. MOC

TABLE 5.1
Examples of Types of Change

Process or System Characteristic	Examples of Change
Hardware	• Equipment added/removed • Equipment configuration • Piping tic-ins • Relief systems • Utility tie-ins • Vessels re-rated
Control Systems	• Conventional system replaced with DCS
Safety Systems	• Safety shutdowns added/removed or new trip points • Safety dumps or purges added, removed, or modified • Fire detection and suppression systems added, removed, or modified
Technology	• New chemicals used or produced, including corrosion inhibitors and other treatment chemicals • Changes in the process, including chemistry and new operating modes • Unit throughput increases/decreases
Siting	• New structures added • Temporary structures or trailers added • Increased occupancy • Changes to emergency vehicle access routes • Increased vehicle traffic • Equipment relocation • Electrical classification changes • New adjacent units • Increases in, or addition of new, vulnerable exposures surrounding the process unit (either on-site or off-site)
Operations	• New or revised procedures • Staffing decreased/increased • Training program modifications • Changes in occupancy patterns/ location of personnel work areas
Management Systems	• Audit frequencies adjusted • Safe Work Practices modified
Other	• Changes in "nonprocess" safeguards (e.g., diminished capabilities from off-site response organizations) • New, or more significant, external events that could impact the process

programs associated with such PHAs may not have been fully implemented or fully effective. Consequently, some modifications to the process may not have received a formal, or a rigorous, MOC review.

5.2.2. P&ID Comparison

P&IDs are typically the focal point of the PHA sessions, and represent one of the more concentrated sources of information about the process. As such, these drawings are a logical place to search for changes, through comparison of the current P&IDs with those used during the prior PHA (if available).

The simplest way to identify modifications to the P&IDs is to check the revision numbers on the current drawings versus the revision numbers on the drawings used during the prior PHA. Another method for identifying changes using the P&IDs is to visually compare each old and corresponding new drawing to identify any new equipment or lines. For every revision after the prior PHA, the nature of the change and other relevant information (e.g., an MOC record identifier) should be documented for consideration during the revalidation sessions.

If there has been a wholesale revision of the P&ID, the above approaches may not be applicable (often, prior revision numbers are consolidated under a "General Update" identification when a P&ID is redrawn). In such circumstances, it may be possible to gather much of the needed information from an archived copy of the last drawing revision prior to the drawing update.

Changes shown on the P&IDs may not necessarily be *controlled* changes. Relevant MOC, PSSR, and/or PHA records should be identified for the changes found on the drawings. If no such documentation can be found, this fact should be documented and the associated change flagged for more detailed review during the revalidation.

5.2.3. Procedure Reviews

A similar comparison can be made between the current versions of operating procedures and the versions in effect at the time of the prior PHA (if archival copies have been retained.) Again, these changes may not have been subject to MOC scrutiny, and a cross-check with the MOC records system should be made.

5.2.4. PHA and Incident Investigation Recommendations

PHA and incident investigation recommendations are the documentation of the intent to make a change, and may be expressed in rather general terms (e.g., "Provide overpressure protection on vessel X."). The specific implementation details are often determined later and some organizations use the MOC system to assess the safety of the proposed means of implementation. Where this is the case, records of implemented recommendations from prior PHA reports(s) and incident investigation reports should be reviewed and compared to the MOC records. If the MOC program is working effectively, the implemented recommendations will have been evaluated under the MOC program. Implemented recommendations that were not so scrutinized should be flagged for more detailed review during the revalidation.

5.3. Undocumented and Uncontrolled Changes

Some changes may not have received a thorough MOC evaluation; for example, sometimes the person responsible simply does not recognize a change as being a change. Additionally, there are some types of change that could have a potential process safety impact, but that are not necessarily covered under the MOC program. Such changes might include:

- Changes in staffing level or key personnel that might impact upon a facility's capability to respond, or the speed of response, to a particular abnormal situation;
- Substitution of contractor operating or maintenance personnel for company personnel;
- Sales of assets within the plant to another company;
- Procedural changes (particularly frequencies of performing tasks); or
- The addition of new occupied buildings, demolition of nearby facilities, or an increase in the number of people in existing buildings near a process, potentially impacting the results of prior siting reviews.

These types of subtle changes should be identified as part of the PHA revalidation effort to ensure that the hazards of the process are appropriately assessed and controlled. The difficulties in identifying and managing subtle changes have been discussed elsewhere (Burk 1992).

Several of the more common sources of information used to identify undocumented and uncontrolled changes are discussed below. Additionally, the *Facility and Process Modification Checklist* provided in Appendix E, may be useful in helping the PHA Team identify changes that may not have been documented under the MOC system.

5.3.1. Interviews of Facility Personnel

Interviews may be considered as part of the revalidation process to verify that PSI is accurate, and to identify changes that might not otherwise be identified through process safety information reviews (e.g., such interviews may be helpful in validating the content of the P&IDs).

Interviews can involve discussions with either the revalidation team members or members of the general population working in or associated with the process unit under review. Identified changes should be recorded, so that they can later be cross-referenced against the MOC records to determine whether they are documented or undocumented changes.

Interviews of the Revalidation Team Members

The team leader can interview members of the revalidation team either individually or as a group during one of the revalidation sessions. When interviewing members of the revalidation team, it is often effective to provide the team members with copies of the *Facility and Process Modification Checklist* prior to the interview so they can research answers to the questions.

Interviews of Unit Personnel

Interviews outside of the revalidation team provide a broader sampling of personnel, involving a cross section of the plant or unit, but with primary focus on:

- Operations (both control room and field personnel);
- Maintenance;
- Inspection;
- Reliability;
- Engineering;
- Environmental;
- Emergency Response; and
- Safety and Health.

Although more time-consuming, interviews of unit personnel can provide advantages over interviews conducted during the revalidation meeting. For example, field interviews can be conducted on a more informal, one-on-one basis. Another advantage is that the persons being interviewed get to point out the changes that have occurred while they are in the unit, and are likely to remember more changes in this context than if they were in a conference room setting. No single individual is likely to remember, or be aware of, all changes and interviews of unit personnel provide a larger source of information to sample.

5.3.2. Maintenance Records

Reviewing each work order or reviewing the maintenance history generated during the five years since the prior PHA would be a daunting task, and should not normally be necessary. However, a random sampling of such records might confirm whether a more detailed review is warranted. Some work order software systems may allow key word searches for indicators of the type of maintenance work performed (e.g., "upgrade," "change," or "modify") to help identify work orders for scrutiny.

If work orders describe activities that are more than just *replacement-in-kind* and no associated MOC documentation can be found, these *changes* should be flagged for more detailed consideration during the revalidation sessions.

5.3.3. Purchase Specifications and Records

Changes in purchase specifications for chemicals, spare parts, equipment, etc., associated with the process are changes that should receive MOC scrutiny. These types of records may also warrant a random inspection to determine if potential problems exist.

5.3.4. Incident Investigation Reports

Reviews of incident investigation reports, with careful attention to the identified causes, may reveal problems with uncontrolled and undocumented changes.

5.3.5. PSM Program Audits

Consideration should be given to reviewing recent PSM program audits, and third-party audits such as those conducted by regulatory agencies. This may help focus search efforts on particular elements of the PSM program that warrant scrutiny.

6

Identifying an Appropriate Revalidation Methodology

While it may be intended that a PHA revalidation be accomplished with less effort and in less time than an initial PHA, the realization of this goal is highly dependent upon careful planning. To develop the revalidation study work plan, it will be necessary to carefully review and integrate the information gathered above. Where are we now (with respect to the prior PHA)? Where do we need to go (considering what has transpired since the prior PHA)?

Much of the preparation conducted up to this point has been directed towards the decision that now must be made; that is, what revalidation approach is appropriate to the situation at hand?

6.1. Revalidation Options

As previously introduced, there are three alternative courses of action from which to choose. The three approaches are briefly summarized below:

6.1.1. Update and Revalidate

Updating and Revalidating is the least laborious option and is reserved for high quality PHAs of processes at facilities where hazards have been effectively managed; e.g., where MOC has been effectively implemented and documented, and few or no significant incidents have occurred. In these situations, assembled documentation can be used to lead the team through the verification and integration of the results from the prior PHA and the hazard reviews associated with MOC/PSSR reviews.

There are two common approaches to *Updating and Revalidating* a PHA:

1. Section-by-section review of the prior PHA documentation; or
2. Change and incident review using supplemental documentation, such as a review worksheet.

The section-by-section review is similar to *Redoing* the PHA, but should entail considerably less time and effort, since the content of the PHA report is only being amended on an "as-needed" basis. Changes are discussed to determine their potential impact on existing event scenarios, or to determine if new scenarios need to be added to the report. Incidents are similarly discussed to determine their significance. Only those scenarios potentially impacted by the changes or the incidents need to be addressed, and necessary changes in the PHA are recorded on the PHA documentation (e.g., the HAZOP worksheets).

Alternatively, the revalidation team may choose to assess the significance of changes and incidents on a more global, rather than section-by-section, basis. The Team, with appropriate reference to the prior PHA, discusses the significance of the changes and incidents and documents their deliberations using a separate analysis table or worksheet. Information recorded on the worksheet could include a description of the change (or incident), its process safety significance, and any risk management decisions or recommendations made by the revalidation team. This worksheet serves as the *Update* to the PHA.

The amount of time needed to perform the *Update and Revalidation* will depend upon the extent and number of changes made to the process, and the number and significance of incidents that have occurred.

Revalidation teams may encounter situations where the prior PHA was of very high quality (i.e., containing no gaps or deficiencies) and where there have been no significant incidents or changes within the subject process. Such circumstances may permit a much simpler revalidation effort and report, limited to affirming the continued validity of the prior PHA.

6.1.2. Retrofit, Update, and Revalidate

This approach is applicable to those situations where the Chapter 4 evaluation has identified gaps or deficiencies that need to be remedied during the revalidation. Correction of these gaps or deficiencies constitutes the *Retrofit* portion of the effort. During the *Retrofit* review, the team should consider whether additional safeguards or other actions are advisable to control process hazards and develop recommendations as appropriate.

There are two common approaches to *Retrofitting* a PHA to address gaps or deficiencies:

1. Section-by-section review of the prior PHA documentation to enhance the treatment of the particular issue(s) associated with the gaps or deficiencies; or
2. Use of supplemental documentation, such as an appropriate topical checklist.

For example, assume that the prior PHA, conducted using the HAZOP methodology, was reviewed in accordance with the guidance given in Chapter 4, and it was determined that the PHA did not provide a "*qualitative evaluation of a range of the possible safety and health effects of failure of controls.*" The revalidation team could review the HAZOP worksheets, and add appropriate consequence rankings, or other qualitative descriptors for the range of effects, to the existing scenarios as appropriate. Many scenarios would likely have been noted as having "no consequences" and would not have to be addressed.

Conceivably, such deficiencies might only exist in certain portions of the PHA report, and only the affected sections would warrant a *Retrofit*. The decision as to whether to *Retrofit* or *Redo* the PHA would be based, in part, upon how pervasive the gaps or deficiencies were throughout the report. For example, assume that the hazards of H_2S in a process were not addressed in the prior PHA. A *Retrofit* might be feasible if the H_2S was present in only a few vessels, whereas a *Redo* might be required if the H_2S was present throughout the process.

Finally, assume that siting had not been addressed within the prior PHA report. The revalidation team might choose to *Retrofit* the PHA via the completion of a *Facility and Stationary Source Siting Checklist*, such as that provided in Appendix F. Alternatively, if the siting issues were more significant than could be addressed with a checklist, the Team might recommend that a separate, stand-alone siting study be completed.

Regardless of the approach used in *Retrofitting* the PHA, the goal of the effort is to provide the additional information that should have been in the prior PHA report at the time it was issued (i.e., to fill the gaps or deficiencies). Making the PHA "whole" in this manner creates a valid foundation upon which the subsequent *Update and Revalidate* effort can build, using the same approaches described in the previous section.

6.1.3. Redo

Redoing a prior PHA may be the most appropriate choice in situations where *significant*

- Gaps or deficiencies exist in the manner in which the prior PHA was conducted;

- Change, particularly uncontrolled change, has occurred in the process or equipment since the prior PHA was conducted; or
- Incidents or near misses have occurred.

There may be circumstances where *Redoing* the PHA may be preferable, not because of gaps or deficiencies in the prior PHA but from a purely logistical standpoint. For example, assume a process has experienced a large number of MOCs since the prior PHA and the company is considering *Updating* its PHA documentation (e.g., HAZOP worksheets) to reflect these changes as part of the revalidation process. *Redoing* the PHA may actually be more time or resource-effective effective than *Updating and Revalidating* the documentation to incorporate each MOC. In other words, reviewing all MOC (and associated PSSR) documents to ensure that the hazards were adequately addressed may require more time than *Redoing* the PHA. Furthermore, if the company intends to continue *Updating and Revalidating* (as opposed to *Redoing*) the PHA for an indefinite period, document retention and revalidation logistics (e.g., determining what was reviewed, what is still valid) could become unmanageable after several revalidations, as more and more MOC documentation is appended to the PHA report. Refer to Chapter 8 for more guidance on documentation issues.

From the Workshop...

The decision to *Redo* a PHA offers the opportunity to reconsider the choice of hazard evaluation technique to be used (e.g., HAZOP, rather than What If). Revalidation Workshop participants debated the potential merits of intentionally alternating techniques from one revalidation to the next. Use of the same technique may be more time- or resource-effective. Conversely, it may contribute to propagating any problems resulting from technique use or selection in the prior PHA. If a technique substitution is made, it should be with the intent of enhancing the strength of the review, and not at the expense of accuracy or thoroughness. Section 4.1.2 discusses technique selection in light of the nature of the process and its hazards.

The *Redo* revalidation method involves conducting an initial PHA of the process using a method that is appropriate for the hazards and complexity of the process. Thus, the requirements and activities for a *Redo* generally parallel those for an initial PHA. For all practical purposes, the effort starts from scratch with a blank sheet of paper (or computer screen). Significant instructional information is available elsewhere on the conduct of PHAs and will not be repeated here (e.g., CCPS G-1).

Revalidation team members should be cognizant of any documentation or analysis shortcomings in the prior PHA and strive to ensure that the *Redo* does not propagate these problems.

One additional consideration might suggest the need for *Redoing* a PHA. Experience to date shows that a well organized and well conducted revalidation can effectively address the changes that occur over a 5-year revalidation cycle. Some companies, concerned that "random noise" might accumulate in the work product after multiple revalidations, have suggested the consideration of *Redoing* a PHA after a certain number of revalidations.

6.2. Selecting the Revalidation Options

Figure 3.1 illustrated the logic underlying the selection of the revalidation approach. The selection of which course of action to use is made considering a number of factors, including:

- Internal and external factors.
- Quality and completeness of the prior PHA
- Operating experience

Decision criteria will commonly be organization-specific and, for this reason, this book only addresses criteria qualitatively. Readers may find other factors, pertinent to their own circumstances, that could also be considered in establishing such criteria.

Internal and External Factors

A number of internal and external factors could impact the course of the revalidation effort. These could include changes in company policies, procedures, or loss prevention goals; new or revised regulatory requirements;

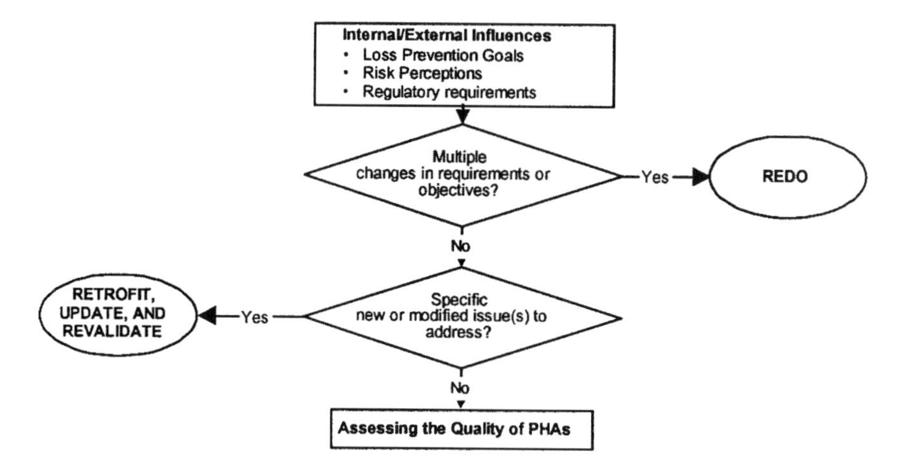

new or revised industry consensus standards; or new information or perspectives about the hazards or risk of the process. Such changes in hazard perspectives might result from new research data or the application of new assessment technologies (e.g., consider the impact that more sophisticated reaction calorimetry techniques have had on reactor safety issues).

If the number or significance of changes in internal or external factors is minor, it may be possible to *Retrofit* the prior PHA to address them, and then proceed to *Update and Revalidate* the PHA. If there have been multiple, or significant changes, it may be more appropriate to *Redo* the PHA.

If there have been no significant changes, or additions to such factors, the next step is to evaluate the prior PHA.

Quality and Completeness of the Prior PHA

PSI needs in support of a quality PHA effort were reviewed in Chapter 3. Guidance for the evaluation of the prior PHA, and checklists specific to the task, were provided in Chapter 4. If the gaps or deficiencies in the prior PHA are few, or minor, they may be addressed via a *Retrofit* of the prior PHA. More substantive problems with the quality or completeness of the prior PHA may require that the PHA be *Redone.*

If there are no significant issues related to the quality or completeness of the prior PHA, the next step is to look at the operating experience in the interim since the prior PHA was conducted.

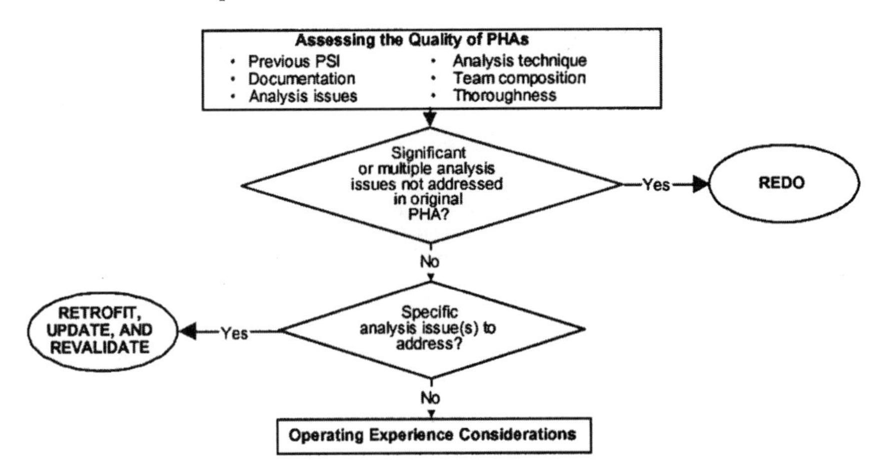

Operating Experience

The frequency and nature of changes made to the process and equipment, and the degree to which this change has been properly controlled are factors to be considered in determining the appropriate revalidation approach.

Additional guidance addressing the identification and evaluation of such changes was provided in Chapter 5.

It is also appropriate to consider the incident history within the process and, where such information is available and relevant, the incident history of similar facilities and processes elsewhere.

Numerous incidents or changes (especially uncontrolled changes) may warrant a *Redo* of the PHA. If the numbers of changes or incidents have been few, but some of these have been significant, it may be possible to reflect these within a *Retrofit, Update and Revalidate.*

Those prior PHAs that pass straight down the decision tree are candidates for an *Update and Revalidate.*

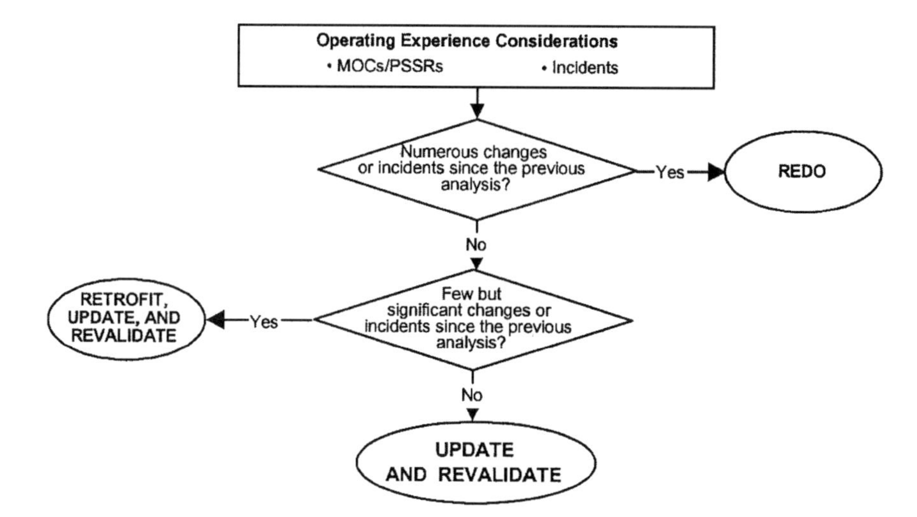

7

Conducting the Revalidation Study Sessions

7.1. Team Training

The training requirements will be specific to the particular needs of the team. However, a number of topics could be considered in determining the training to be provided.

- ***Provide a process overview and conduct a facility tour.*** While most team members will likely have at least a basic familiarity with the process and equipment, it is often helpful to begin the team training by reviewing the process and making a brief tour of the facility. This helps ensure a common familiarity with the process and may help identify and correct any misconceptions that team members might have as to the scope of the review.
- ***Describe the conduct of a PHA and review what it is intended to accomplish.*** Team members who have not previously participated in a PHA need to understand the purpose of the PHA (see Chapter 1 of this book). Some personnel may not be accustomed to the critical analysis process used in a PHA, or may be unfamiliar with the cause/consequence/protections logic of the PHA process. These concepts should be reviewed with the team members.
- ***Review the purpose of a revalidation.*** The team leader should consider reviewing the reasons for the revalidation and what it is intended to accomplish, emphasizing the need for a thorough but efficient review (see Chapter 2 of this book).
- ***Review the selected revalidation methodology.*** The team will need a basic understanding of the methodology that will be used. A skilled facilitator will typically lead the team through a period of "on-the-job training" during the initial revalidation session(s), but this training

period will be more effective if prefaced by a brief introduction to the methodology.

- ***Review the record of changes.*** If the decision has been made to *Update and Revalidate* or *Retrofit, Update and Revalidate* the PHA, the Team members should be made familiar with the reason for, and the contents of, the change record prepared ahead of the sessions. It is not intended that the change record be reviewed change-by-change at this time, since it is designed to be used throughout the course of the revalidation. However, team members need to know how to integrate the information into their deliberations.

7.2. Application of Revalidation Methodology

Chapter 6 detailed the three alternative courses of action (*Update and Revalidate; Retrofit, Update and Revalidate;* and *Redo*) and described the different means of implementing them.

7.3. Special Topics

7.3.1. Staying Productive

It should be remembered that a revalidation session is still a PHA session, albeit perhaps slightly more structured. The commonsense learnings that trained PHA facilitators acquire on-the-job are as applicable to the efficient conduct of a PHA revalidation as they are to the conduct of an initial PHA.

There are, however, some points to be raised with respect to the unique nature of the revalidation session that can be considered when attempting to keep the sessions as productive as possible.

As may be evident by now, a significant portion of the revalidation effort is spent reviewing past work products and supporting documentation. This provides greater opportunity for customizing the attendance in a particular session to meet the particular topic of the session. For example, if the session topic is a review of maintenance work orders to determine the significance of any changes identified, the quorum for this session might not include the control engineer responsible for the distributed control system, or the laboratory supervisor. Or, if the focus of the session is to evaluate changes in the operating procedures, involvement of the maintenance supervisor, or inspection supervisor might be superfluous.

Resources are more wisely used, and personnel interest and motivation is more easily maintained, if the attendance in a particular session is limited to those personnel germane to the particular topic.

There is a significant amount of preparatory work involved in locating, collecting, and preparing information for use in the revalidation sessions. It is incumbent that the study leader, or designee, does the needed legwork ahead of the meeting so that this information can be accessed and used as efficiently as possible. For example, P&ID revisions can be cross-referenced to PHA study nodes, and to MOC and PSSR documentation *before the sessions begin*. Sessions will be more productive, and less onerous for participants, if perusal of records and shuffling of papers can be minimized by proper preparation ahead of time.

7.3.2. Facility (or Stationary Source) Siting

Where detailed siting analyses have been conducted, study documentation and personnel with the expertise to interpret the studies should be available for the revalidation sessions. Not all situations, however, require that the siting issue be addressed via quantitative studies. Many companies have developed checklists to aid PHA and revalidation teams in addressing siting issues.

Siting checklists often address topics such as:

- Unit layout and spacing between equipment;
- Location of large inventories of hazardous materials;
- Location and construction of the control room and other occupied buildings (e.g., engineering, maintenance, laboratory, administration);
- Location of motor control centers and other electrical area classification issues;
- Location of other likely sources of ignition;
- Location of the facility relative to on-site and off-site neighbors;
- Location of firewater mains and backup;
- Location and adequacy of drains, spill basins, dikes, and sewers; and
- Location of emergency stations (showers, respirators, personal protective equipment).

An example *Facility and Stationary Source Siting Checklist* is provided in Appendix F.

7.3.3. Human Factors

Similarly, many companies have found the use of checklists advantageous in helping teams address human factor issues within PHA and PHA revalidations.

Human factors checklists often include topics such as:

- Equipment identification;
- Accessibility and availability of controls and equipment;
- Feedback/displays of controls;
- Personnel workload/stress;
- Training;
- Procedures; and
- Housekeeping.

From the Workshop...

One of the needs by the Revalidation Workshop participants was that for more guidance, and perhaps better tools, for addressing human factors in PHAs.

An example *Human Factors Checklist* is provided in Appendix G.

7.3.4. Wrap-Up Discussions

Some revalidation team leaders find it effective, after all other items are discussed, to guide the revalidation team through a period of general discussion to ensure that no significant topics were overlooked. These wrap-up discussions can be used to identify global hazards that may have been missed during previous discussions, such as loss of instrument air to the entire process unit or section, or the impact of external events such as earthquakes or tornadoes. Additionally, the wrap-up discussion can be used to ensure that the PHA revalidation complies with applicable company or regulatory requirements by specifically reviewing and documenting discussions on required considerations. Such considerations could include human factors, siting, and previous incidents (while these topics should have been addressed earlier in the PHA, some leaders like to revisit them at the conclusion of the study to ensure their comprehensive treatment). Finally, the wrap-up discussion may be used to document discussion of issues that are common to multiple study sections, such as corrosion monitoring programs or relief header/flare sizing. Some areas of discussion may be enhanced if specialist personnel with particular expertise participate.

Readers may wish to develop a checklist, specific to their particular needs, for use guiding these discussions. The following list suggests topics that might be considered during such wrap-up discussions.

- Safety and fire
  —Fire protection systems
  —Area electrical classification
  —Industrial hygiene
- Emergency response
  —Community Emergency Response Plan
  —In-plant Emergency Response Plan

- Procedures
- Loss of utilities
- Siting
- Previous incidents
- Human Factors
- Mechanical Integrity
  - —Testing and inspection
  - —Training
- Maintenance
- External events
- Recommendations
  - —From previous PHA Studies
  - —From incident investigations
- Other company or local requirements

8

Documenting the Revalidation Study

A clear, concise, thorough PHA revalidation report is essential to the retention and communication of the PHA results. Experience shows that one of the most difficult obstacles for a revalidation team results from inadequate documentation of the prior PHA. It is likely that most PHA revalidation teams will gain, during the course of the PHA revalidation, a greater appreciation of the importance of clear, accurate documentation and respond accordingly.

Much of the documentation for a PHA revalidation will parallel that for an initial PHA. However, some approaches to revalidation, as suggested in this book, use a number of screening forms and checklists. Those choosing to use such forms and checklists should consider including them in the revalidation documentation so that the basis for key decisions (e.g., the basis for the particular revalidation option chosen) can be clearly communicated to the next revalidation team, five years hence.

The PHA revalidation report format and content will often be prescribed by company requirements. This chapter provides suggested documentation practices for the PHA revalidation and associated records that could be used in lieu of more specific company requirements.

8.1. Documentation Approaches

The two basic approaches to documenting the PHA are revalidation "evergreen" documentation and "basic" documentation.

With "evergreen" documentation, a new PHA report, similar in format and content to the prior PHA report, is prepared. This single document would:

1. Identify the process unit examined;
2. List meeting participants;
3. List documents examined (e.g., previous PHA report, P&IDs, incidents since the previous PHA);
4. Describe the revalidation approach and the rationale for its selection; and
5. Document the analysis results and findings.

From the Workshop...

Revalidation Workshop participants noted that one common problem with documenting PHA revalidations occurred when changing the software used to document the PHA. Lack of "backwards compatibility" often resulted in the need to retype the prior PHA worksheets for the revalidation effort. This can have a significant impact on the time required to prepare for, or complete the documentation of, the revalidation.

Appendices to the report could contain the PHA technique worksheets (e.g., HAZOP worksheets) and other supporting documentation such as the *Change Summary Worksheet* or any appropriate checklists used by the revalidation team.

This documentation style is more commonly used when the PHA worksheets are revised during the revalidation. "Evergreen" documentation should result in a report that:

- Integrates and presents the requisite information in a format that looks familiar to users;
- Simplifies revalidation in the future;
- Demonstrates that all analysis issues were addressed;
- Can be used to support other PSM activities (e.g., mechanical integrity, training); and
- Accurately represents the process configuration, hazards, and applicable controls in one document.

The more "basic" form of documentation involves compiling the requisite individual documents such as the prior PHA, completed MOC forms, incident investigation reports, etc., along with any worksheets or checklists completed during the revalidation. These documents are then linked together by a summary cover report that summarizes the results of the revalidation.

This simplified approach may make it more difficult for some to use the report, since it does not provide the information in an integrated format. Consequently, the "basic" form of documentation is commonly limited to those situations for which relatively little *Update* has been required in the *Update and Revalidate* (e.g., where there have been few significant changes or incidents to reflect within the revalidation).

8.2. Report and Its Content

The following is a suggested table of contents for the revalidation report.

- Summary Report
  —Purpose and introduction
  —Description of process unit/section
  —Study dates and member attendance
  —List of team members indicating their job function (e.g., operator, engineer)
  —Description and substantiation of revalidation method
  —Summary of recommendations
- Reference Information
  —List of P&IDs, including revision number or date of last revision
  —Copy of P&IDs
  —List of study sections (if applicable)
  —List of PSI referenced in study
- Study Worksheets
  —Worksheets for each drawing reviewed (if applicable)
  —Worksheets for each section reviewed (if applicable)
- Other Completed Exhibits (if used)
  —Essential Criteria Checklist (see Appendix B)
  —PHA Quality and Completeness Checklist (see Appendix C)
  —Change Summary Worksheet (see Appendix D)
  —Facility and Process Modification Checklist (see Appendix E)
  —Facility and Stationary Source Siting Checklist (see Appendix F)
  —Human Factors Checklist (see Appendix G)
  —Wrap-up Discussion Checklist (see Section 7.3.4)

Additional information on documentation for various hazard evaluation techniques is provided in the CCPS *HEP Guidelines* and in the CCPS *Guidelines for Process Safety Documentation* (CCPS G-27).

8.3. Recommendation Follow-Up

As with any PHA recommendation, the recommendations coming out of the revalidation study should be resolved in a timely manner. Experience indicates that explicit, clear documentation of responsibilities is essential to the resolution of recommendations. In general, organizations should consider documenting the following with respect to recommendation follow-up:

- What is to be done;
- Schedule for completion;
- Who is responsible for follow-up;
- Periodic status reports until the recommendation is resolved;
- Manner in which the recommendation was resolved;
- Date the recommendation was resolved; and
- List of employees informed of the recommendation and its resolution.

8.4. Records Retention and Distribution

The PHA report and any PHA revalidation reports are valuable sources of information for those having a responsible role in controlling the hazards of the process. Furthermore, these documents can be valuable training tools. Companies should consider implementing effective ways of ensuring that this information is conveniently accessible.

Additionally, it is important that this valuable information be protected. Company policies or procedures often establish retention practices for the PHA-related records. Specific requirements are imposed upon those facilities covered by the OSHA PSM or EPA RMP regulations; the PHA and PHA revalidation reports, as well as documentation of the resolution of recommendations from these reports, must be maintained for the life of the process. Covered facilities should ensure that archival copies of all versions of the PHA documentation are maintained, even when the "evergreen" documentation option is used to update the prior PHA report.

Appendix A

Federal Regulatory Requirements

The Clean Air Act (CAA) Amendments of 1990 required that both OSHA and EPA develop regulations intended to prevent accidental releases of highly hazardous materials. In February 1992 OSHA issued a chemical accident prevention regulation, its Process Safety Management (PSM) Standard (OSHA 1992). The PSM Standard applies to those processes (i.e., "covered processes") containing more than a specified threshold quantity of a listed hazardous material.

Included in the accident prevention provisions of the PSM Standard is the requirement to conduct process hazard analyses for covered processes. OSHA provides general requirements for the conduct of the PHA, addressing issues such as scheduling, team make-up, techniques allowed to be used, content and scope of the PHA, and documentation requirements. The requirement pertinent to the subject of this book is the requirement that the PHA be updated and revalidated at an interval not to exceed five years. The PHA requirements from the PSM Standard are reproduced verbatim in Figure A-1.

EPA similarly issued a chemical accident prevention regulation, the Risk Management Program (RMP) Rule, in June 1996 (EPA 1996). Unlike OSHA's "one size fits all" approach, EPA established three different program levels to address facilities with different levels of perceived hazard potential. The accident prevention requirements for EPA's most rigorous level, Program Level 3, generally parallel those of OSHA PSM, including the requirements for conducting PHAs and subsequently updating and revalidating them every five years. These requirements are reproduced in Figure A-2.

While the OSHA PSM and the EPA RMP Program Level 3 requirements are very similar, there is an important difference in regulatory focus. OSHA has responsibility solely for the facility worker, while EPA has responsibility for off-site populations and the environment. These different regulatory responsibilities are reflected in the regulatory requirements. For example, a PHA conducted to meet the OSHA requirements must address:

> *A qualitative evaluation of a range of the possible safety and health effects of failure of controls on employees in the workplace.*

In contrast, a PHA conducted to meet the EPA requirements must more broadly address:

> *A qualitative evaluation of a range of the possible safety and health effects of failure of controls.*

In the first case, the focus is restricted to the effects on workers; in the latter case, the focus is expanded to include the effects beyond the fenceline. (See the sidebar below.)

For its less rigorous Program Level 2 prevention program, EPA requires a hazard review, rather than a PHA. As shown in Figure A-3, the requirements for a hazard review are much less detailed than those for a PHA. However, updating and revalidating the hazard review on a five-year period is still required.

There are no accident prevention program requirements for Program Level 1 facilities.

The RMP Rule, in §68.190, specifies a number of circumstances under which the risk management plan required to be filed under the Rule must be

Different Sides of the Fenceline

In their joint Memorandum of Understanding, OSHA and EPA acknowledge their different regulatory mandates:

"The goal of this strategy for coordinated implementation is to ensure effective integration of both EPA's and OSHA's chemical accident prevention activities while recognizing each Agency's historical expertise and mission: OSHA's in protecting worker safety and health and EPA's in protecting public health and the environment."

(e) *Process hazard analysis.*

(1) The employer shall perform an initial process hazard analysis (hazard evaluation) on processes covered by this Standard. The process hazard analysis shall be appropriate to the complexity of the process and shall identify, evaluate, and control the hazards involved in the process. Employers shall determine and document the priority order for conducting process hazard analyses based on a rationale which includes such considerations as extent of the process hazards, number of potentially affected employees, age of the process, and operating history of the process. The process hazard analysis shall be conducted as soon as possible, but not later than the following schedule:

- (i) No less than 25 percent of the initial process hazards analyses shall be completed by May 26, 1994;
- (ii) No less than 50 percent of the initial process hazards analyses shall be completed by May 26, 1995;
- (iii) No less than 75 percent of theinitial process hazards analyses shall be completed by May 26, 1996;
- (iv) All initial process hazards analyses shall be completed by May 26, 1997.
- (v) Process hazards analyses completed after May 26, 1987 which meet the requirements of this paragraph are acceptable as initial process hazards analyses. These process hazard analyses shall be updated and revalidated, based on their completion date, in accordance with paragraph (e)(6) of this section.

(2) The employer shall use one or more of the following methodologies that are appropriate to determine and evaluate the hazards of the process being analyzed.

- (i) What-If;
- (ii) Checklist;
- (iii) What-If/Checklist;
- (iv) Hazard and Operability Study (HAZOP):
- (v) Failure Mode and Effects Analysis (FMEA);
- (vi) Fault Tree Analysis; or
- (vii) An appropriate equivalent methodology.

(3) The process hazard analysis shall address:

- (i) The hazards of the process;
- (ii) The identification of any previous incident which had a likely potential for catastrophic consequences in the workplace;
- (iii) Engineering and administrative controls applicable to the hazards and their interrelationships such as appropriate application of detection methodologies to provide early warning of releases. (Acceptable detection methods might include process monitoring and control instrumentation with alarms, and detection hardware such as hydrocarbon sensors.);
- (iv) Consequences of failure of engineering and administrative controls;
- (v) Facility siting;
- (vi) Human factors; and
- (vii) A qualitative evaluation of a range of the possible safety and health effects of failure of controls on employees in the workplace.

(4) The process hazard analysis shall be performed by a team with expertise in engineering and process operations, and the team shall include at least one employee who has experience and knowledge specific to the process being evaluated. Also, one member of the team must be knowledgeable in the specific process hazard analysis methodology being used.

(5) The employer shall establish a system to promptly address the team's findings and recommendations; assure that the recommendations are resolved in a timely manner and that the resolution is documented; document what actions are to be taken; complete actions as soon as possible; develop a written schedule of when these actions are to be completed; communicate the actions to operating, maintenance and other employees whose work assignments are in the process and who may be affected by the recommendations or actions.

(6) At least every five (5) years after the completion of the initial process hazard analysis, the process hazard analysis shall be updated and revalidated by a team meeting the requirements in paragraph (e)(4) of this section, to assure that the process hazard analysis is consistent with the current process.

(7) Employers shall retain process hazards analyses and updates or revalidations for each process covered by this section, as well as the documented resolution of recommendations described in paragraph (e)(5) of this section for the life of the process.

Figure A-1. PHA Requirements from the OSHA PSM Standard (29 CFR 1910.119)

§68.67 Process hazard analysis.

(a) The owner or operator shall perform an initial process hazard analysis (hazard evaluation) on processes covered by this part. The process hazard analysis shall be appropriate to the complexity of the process and shall identify, evaluate, and control the hazards involved in the process. The owner or operator shall determine and document the priority order for conducting process hazard analyses based on a rationale which includes such considerations as extent of the process hazards, number of potentially affected employees, age of the process, and operating history of the process. The process hazard analysis shall be conducted as soon as possible, but not later than June 21, 1999. Process hazards analyses completed to comply with 29 CFR 1910.119(e) are acceptable as initial process hazards analyses. These process hazard analyses shall be updated and revalidated, based on their completion date.

(b) The owner or operator shall use one or more of the following methodologies that are

(1) What-If;
(2) Checklist;
(3) What-If/Checklist;
(4) Hazard and Operability Study (HAZOP);
(5) Failure Mode and Effects Analysis (FMEA);
(6) Fault Tree Analysis; or
(7) An appropriate equivalent methodology.

(c) The process hazard analysis shall address:

(1) The hazards of the process;
(2) The identification of any previous incident which had a likely potential for catastrophic consequences.
(3) Engineering and administrative controls applicable to the hazards and their interrelationships such as appropriate application of detection methodologies to provide early warning of releases. (Acceptable detection methods might include process monitoring and control instrumentation with alarms, and detection hardware such as hydrocarbon sensors.);
(4) Consequences of failure of engineering and administrative controls;
(5) Stationary source siting;
(6) Human factors; and
(7) A qualitative evaluation of a range of the possible safety and health effects of failure of controls.

(d) The process hazard analysis shall be performed by a team with expertise in engineering and process operations, and the team shall include at least one employee who has experience and knowledge specific to the process being evaluated. Also, one member of the team must be knowledgeable in the specific process hazard analysis methodology being used.

(e) The owner or operator shall establish a system to promptly address the team's findings and recommendations; assure that the recommendations are resolved in a timely manner and that the resolution is documented; document what actions are to be taken; complete actions as soon as possible; develop a written schedule of when these actions are to be completed; communicate the actions to operating, maintenance and other employees whose work assignments are in the process and who may be affected by the recommendations or actions.

(f) At least every five (5) years after the completion of the initial process hazard analysis, the process hazard analysis shall be updated and revalidated by a team meeting the requirements in paragraph (d) of this section, to assure that the process hazard analysis is consistent with the current process. Updated and revalidated process hazard analyses completed to comply with 29 CFR 1910.119(e) are acceptable to meet the requirements of this paragraph.

(g) The owner or operator shall retain process hazards analyses and updates or revalidations for each process covered by this section, as well as the documented resolution of recommendations described in paragraph (e) of this section for the life of the process.

Figure A-2. PHA Requirements from the EPA RMP Rule (40 CFR Part 68) for Program Level 3 Processes

§68.50 Hazard review.

(a) The owner or operator shall conduct a review of the hazards associated with the regulated substances, process, and procedures. The review shall identify the following:

(1) The hazards associated with the process and regulated substances;
(2) Opportunities for equipment malfunctions or human errors that could cause an accidental release;
(3) The safeguards used or needed to control the hazards or prevent equipment malfunction or human error; and
(4) Any steps used or needed to detect or monitor releases.

(b) The owner or operator may use checklists developed by persons or organizations knowledgeable about the process and equipment as a guide to conducting the review. For processes designed to meet industry Standards or Federal or state design rules, the hazard review shall, by inspecting all equipment, determine whether the process is designed, fabricated, and operated in accordance with the applicable Standards or rules.

(c) The owner or operator shall document the results of the review and ensure that problems identified are resolved in a timely manner.

(d) The review shall be updated at least once every five years. The owner or operator shall also conduct reviews whenever a major change in the process occurs; all issues identified in the review shall be resolved before startup of the changed process.

Figure A-3. Hazard Review Requirements from EPA RMP Rule (40 CFR Part 68)for Program Level 2 Processes

updated. Some of these regulatory triggers (e.g., addition of a newly regulated substance to the EPA list of chemicals, or an increase in the inventory of a regulated substance in the process) might prompt the facility to consider revalidating the associated PHA. The reader may wish to refer to the Rule for additional details.

Appendix B

Essential Criteria Checklist*

OBJECTIVE: To evaluate the prior PHA against essential criteria established by company and regulatory requirements. A "No" answer to any of the following questions is singularly sufficient to trigger a *Redo* of the PHA.

CRITERIA	Yes/No?
1. Did the qualifications of the PHA leader/facilitator meet all company and regulatory requirements; i.e., was the leader/facilitator trained, experienced and competent in the PHA method used?	
2. Did the PHA Team make-up/qualifications meet all company and regulatory requirements; i.e., did team include, as a minimum: • a stable team membership roster with minimal substitution of team members during the course of the review; • operations team member(s) with adequate process and equipment knowledge including recent hands-on operating experience; and • an engineer with industry and specific process experience?	
3. Was the prescribed PHA method appropriate for the complexity of the process studied; i.e., was the PHA method justly rigorous for addressing all potential hazards of the process?	
4. Was the PHA method used from the approved list below or was an appropriate equivalent methodology used? • What-If; • Checklist; • What-If/Checklist; • Hazard and Operability Study (HAZOP); • Failure Mode and Effects Analysis (FMEA); or • Fault Tree Analysis.	

* This checklist is provided for illustrative purposes only. Readers may wish to develop such a checklist specific to their own situation and needs.

CRITERIA	Yes/No?
5. Is the PHA documentation sufficient, or can sufficient documentation be reconstructed, to: • verify PHA leader/facilitator and team qualifications; • indicate PHA team meeting dates; • verify that previous incidents, including those with the potential for catastrophic consequences, were reviewed by the PHA team; • verify the PSI utilized by the team was adequate to ensure a thorough study; • review the PHA team's findings including daily worksheets, engineering and administrative controls cited (safeguards), the failure of controls (consequences), and their causes; • verify that siting was addressed; • verify that human factors were addressed; and • verify that a qualitative evaluation of a range of the possible safety and health effects was conducted (risk ranking or some other documented technique)? Note: if only a few of these bullets are an issue, they are probably "fixable," however, if a preponderance of them are an issue, then there is probably a sufficient lack of data to warrant a "No" answer.	

Appendix C

PHA Quality and Completeness Checklist*

OBJECTIVE: To evaluate the prior PHA against quality and completeness criteria established by company and regulatory requirements. A "No" response to any item requires that the issue be adequately considered during the revalidation PHA study sessions.

Verification Questions	Yes, No, Maybe	Evidence of Compliance	Comments on Adequacy of Compliance
A. ACCESS TO AND USE OF PROCESS SAFETY INFORMATION			
1. Is there evidence that the PSI was complete and available for the PHA team to use (or were recommendations generated to complete the PSI)?			
2. Is there evidence that the PSI contained up-to-date P&IDs?			
3. Is there evidence that unit procedures were available?			
4. Is there evidence that the PSI contained design temperatures and pressures for major equipment (e.g., vessels, heat exchangers, pumps)?			
5. Is there evidence that the PSI contained relief device design and setpoint data?			
6. Is there evidence that the PSI contained ventilation system design data?			

* This checklist is provided for illustrative purposes only. Readers may wish to develop such a checklist specific to their own situation and needs.

Verification Questions	Yes, No, Maybe	Evidence of Compliance	Comments on Adequacy of Compliance
7. Is there evidence that the PSI contained electrical classification drawings and information?			
8. Is there evidence that the PSI contained a list of materials of construction, flange ratings, and temperature and pressure limits for piping (if this information was not on the P&IDs)?			
B. HAZARDS OF THE PROCESS			
1. Are all the covered chemicals documented in the PHA report? (flammable, reactive and toxic)			
2. Are all pertinent hazards (fire, explosions, BLEVEs, toxicity, chemical burn, asphyxiation, etc.) associated with releases of all covered chemicals (highly hazardous chemicals, HHCs) in the process addressed in the PHA?			
3. Is all equipment containing HHCs, or that could contain HHCs, addressed in the PHA? (Compare the analysis nodes or sections to process flow diagrams and/or P&IDs)			
4. Is contamination of the process chemicals addressed in the PHA?			
5. Is loss of utilities addressed in the PHA?			
6. Is the unit flare header (and KO drum if one is provided) addressed in the PHA?			
7. If the flare header is not covered in the PHA, will it be addressed in a separate PHA?			
8. Were the MSDSs reviewed by the PHA team?			
C. PREVIOUS INCIDENTS WITH POTENTIAL FOR CATASTROPHIC CONSEQUENCES			
1. Is there evidence that the PHA team reviewed/discussed previous incidents associated with the unit? Evidence may include: • a list or copy of incident reports • a worksheet documenting incident discussions.			

Verification Questions	Yes, No, Maybe	Evidence of Compliance	Comments on Adequacy of Compliance
2. Is there evidence that near-misses were included in the discussion of previous incidents?			
D. APPLICABLE ENGINEERING AND ADMINISTRATIVE CONTROLS			
1. Does the PHA report document the engineering and administrative controls applicable to the process hazards identified?			
2. Is there evidence that the PHA team considered or addressed process instrumentation, alarms, interlocks, programmable logic controllers, DCS?			
3. Is there evidence that the PHA team considered or addressed relief valves?			
4. Is there evidence that the PHA team considered or addressed isolation valves and emergency shutdown systems?			
5. Is there evidence that the PHA team considered or addressed release detection and control systems such as H_2S, hydrocarbon, and HF monitors? (also see Siting)			
6. Is there evidence that the PHA team considered or addressed release detection and control systems such as dikes or other containment structures around large liquid inventories?			
7. Is there evidence that the PHA team considered or addressed flares and neutralizing systems?			
8. Is there evidence that the PHA team considered or addressed PPE (SCBA, protective clothing, showers, respirators, etc.)? (also see Siting)			
9. Is there evidence that the PHA team considered or addressed fire suppression and extinguishing systems? (also see Siting)			
10. Is there evidence that the PHA team considered or addressed emergency response plans? (also see Siting)			

Verification Questions	Yes, No, Maybe	Evidence of Compliance	Comments on Adequacy of Compliance
11. Is there evidence that the PHA team considered or addressed redundancy of equipment (pumps, instruments, uninterruptible power supplies for critical components, etc.)?			
12. Is there evidence that the PHA team considered or addressed maintenance/testing program for equipment, instrumentation/alarms, etc.?			
13. Is there evidence that the PHA team considered or addressed control room construction and applicable safeguards against releases (detectors, pressurization and alarms, intake elevation, blast resistance, separation distance, PPE, etc.)? (also see Siting)			
E. CONSEQUENCES OF FAILURE OF ENGINEERING AND ADMINISTRATIVE CONTROLS			
1. Is there evidence that the worst-case consequences were documented for postulated accident scenarios?			
2. Did the PHA team optimistically (and improperly) assume that engineering controls (e.g., relief valves, shutdown systems, alarms, trips, interlocks) always work when demanded to mitigate or stop an accident scenario?			
3. Did the PHA team rely on safe work procedures or other administrative controls (e.g., operating procedures, lockout-tagout procedures, work permits, management of change) to eliminate or reduce the severity of accident scenarios that could be serious if administrative controls are ignored or violated?			
4. Are administrative and engineering controls documented in the PHA (safeguards or defenses in PHA tables, list of generic safeguards, etc.)?			
5. Is there evidence that the PHA team considered electrical classification for the unit?			

Verification Questions	Yes, No, Maybe	Evidence of Compliance	Comments on Adequacy of Compliance
F. SITING			
1. Is there evidence that the PHA team considered siting issues by performing a checklist analysis?			
2. Is there evidence that the PHA team considered siting issues by performing a walkthrough of the process unit?			
3. Is there evidence that the PHA team considered siting issues by performing a plot plan review of equipment and buildings in or near the unit?			
4. Is there evidence that the PHA team considered siting issues by preparing documentation of buildings, distances from flammable/toxic threats, design criteria, and potential (qualitative) consequences to both on-site and off-site personnel caused by accidents in the unit?			
5. Is there evidence that the PHA team considered siting issues by preparing recommendations pertaining to siting or location of buildings (separation distances, location of air intakes, ignition sources, drainage, etc.)?			
G. HUMAN FACTORS			
1. Is there evidence that the PHA team considered human factors issues by performing a checklist analysis?			
2. Is there evidence that the PHA team considered human factors issues by performing a walkthrough of the process unit?			
3. Is there evidence that the PHA team considered human factors issues by performing a procedures review with a focus on reducing the potential for human errors (ambiguous steps, missing steps, wrong steps, etc.)?			

Verification Questions	Yes, No, Maybe	Evidence of Compliance	Comments on Adequacy of Compliance
4. Is there evidence that the PHA team considered human factors issues by incorporating human errors as causes of potential accident scenarios identified and evaluated by the team?			
5. Is there evidence that the PHA team considered human factors issues by identifying labeling or other ergonomic issues?			
6. Is there evidence that the PHA team considered human factors issues by identifying operator staffing/workload/ schedule issues?			
7. Is there evidence that the PHA team considered human factors issues by preparing recommendations addressing ways to reduce human errors?			
H. EVALUATION OF QUALITATIVE RANGE OF POSSIBLE HEALTH AND SAFETY EFFECTS OF FAILURE OF CONTROLS			
1. Is there evidence that the PHA team qualitatively evaluated the range of effects for accident scenarios postulated (e.g., impact area, references to MSDSs, health/safety effects possibly identified as part of the hazards of the process)?			
I. CONDUCT AND DOCUMENTATION OF THE PHA			
1. For initial studies where the study team did not have actual operating experience with the process or a similar process (e.g., projects for de-bottlenecking or major modifications), are those sections/nodes for which the team lacked this experience currently operating using the design limits assumed in the initial PHA?			
2. Is the PHA method applied consistently throughout the prior study? Evidence includes: • a description of the PHA method in the report. • Proper documentation such as "No new causes discovered" or "No New Issues (NNI)" when the team could not identify unique causes under a deviation.			

Verification Questions	Yes, No, Maybe	Evidence of Compliance	Comments on Adequacy of Compliance
3. Does the PHA report indicate that all modes of operation were considered by the team using discussion guidewords and descriptions of these guidewords? Is there documentation to support that the PHA addressed the unit hazards during nonroutine modes of operation (e.g., startup, shutdown, emergency, maintenance, sampling)?			
4. If a what-if methodology was used: • Were the analysis sections small enough to identify hazards? • Was a checklist used to formulate questions? • Are all appropriate questions documented for each process section?			
5. If an FMEA was used: • Is the system boundary clearly defined? • Are concurrent failures analyzed? • Are all components shown and numbered?			
6. If a HAZOP analysis was used: • Are all nodes (or process sections) documented? • Are all applicable deviations in each node documented even if there were no consequences of interest? • Are loss of containment deviations considered?			
7. Is there evidence that the PHA considered human errors (operations, maintenance, design, procurement, etc.) as causes or initiating events of accidents?			
8. Is there evidence that the PHA considered equipment failures (fatigue, vibration, defective material, overpressure, weld failure, corrosion, erosion, etc.) as causes or initiating events of accidents even though the equipment may have safety or shutdown systems specifically designed to prevent the postulated failure?			

Verification Questions	Yes, No, Maybe	Evidence of Compliance	Comments on Adequacy of Compliance
9. Does the documentation for closure of safety recommendations meet the company requirements?			
10. Did a review of all operability recommendations from the prior PHA confirm there were no safety implications to the scenarios?			

Appendix D

Example Change Summary Worksheet

OBJECTIVE: To record both controlled and uncontrolled changes identified during documentation reviews and interviews. Changes will be reviewed during PHA revalidation sessions.

DRAWING NUMBER	IDENTIFIED CHANGE	SOURCE	REFERENCE/ COMMENTS	ACTION FOR REVALIDATION

Appendix E

Facility and Process Modification Checklist*

OBJECTIVE: To serve as an aid in identifying and recording changes that may have occurred in the process or facility. Changes will be reviewed during PHA revalidation sessions.

Facility Modifications	Response	Recommendations
1. Have buildings, trailers, etc., been constructed or relocated since the previous PHA such that they could be affected by a release?		
2. Have staffing levels changed since the previous PHA such that capabilities to quickly respond to emergency situations have diminished?		
3. Have traffic patterns (e.g., new rail spur, new chemical truck routing) changed since the previous PHA (e.g., could the process be affected by a new release source, can emergency responders reach the process easily, are external impacts more likely)?		
4. Has the operator training program remained substantially the same or been enhanced since the previous PHA?		
5. Have maintenance practices remained substantially the same or been enhanced since the previous PHA?		

* This checklist has been reproduced from Smith and Whittle (2000) and is provided for illustrative purposes only. Readers may wish to develop such a checklist specific to their own situation and needs.

Facility Modifications	Response	Recommendations
6. Have inspection and test practices remained substantially the same or been enhanced since the previous PHA, particularly for standby safety systems (e.g., interlocks, relief valves)?		
7. Are new hazards present due to new chemicals being used?		
8. Are new industry Standards or guidelines that apply to relief system design for equipment in this process being implemented or evaluated?		
9. Have atmospheric detectors for hazardous chemicals been relocated, removed, abandoned, or disabled?		
10. Have control displays changed such that information necessary to diagnose and respond to upset conditions is not readily accessible?		
11. Are new industry Standards or guidelines that apply to inspection and testing of equipment in the process being implemented or evaluated?		
12. Are there new ignition sources near this process?		
13. Has the fuel changed for fired heaters (e.g., furnaces, reboilers)?		
14. Have heating or cooling media been changed?		
15. Has the process sewer system changed (e.g., now a closed system)?		
16. Have normal inventories of hazardous chemicals changed such that the consequences of a release have been significantly altered?		
17. Have operator communication systems changed?		
18. Has spill containment equipment or systems changed?		
19. Have process water treatment systems changed?		
20. Are there new units or equipment nearby that could affect this process?		
21. Are equipment and piping system labels being maintained?		
22. Are color coding systems for piping and components being maintained?		

Facility Modifications	Response	Recommendations
23. Is a higher pressure utility being used to prepare equipment for maintenance?		
24. Do equipment cleaning procedures require the use of a new chemical?		
25. Are new utilities being used to prepare equipment for maintenance?		
26. Have contaminant levels or contaminants in raw materials changed such that new hazardous deposits could be formed?		
27. Have emergency alarm or notification systems changed?		
28. Are safe work practice requirements substantially the same or more rigorous? • confined space entry • entry into process area • lockout/tagout • "linebreaking" • lifting over process equipment • etc.		
29. Have fire detection or suppression systems been modified such that they require a different operator response?		
30. Has the area electrical classification changed such that some equipment is not properly rated for its service?		
31. Is new/modified equipment properly rated for its area electrical classification?		
32. Have facility modifications made alarms difficult to see or hear?		
33. Have facility modifications increased the number of locations that should have PPE available?		
34. Have facility modifications introduced a need to install or relocate existing safety shower/eyewash stations?		
35. Have facility modifications resulted in the potential to overwhelm an operator with alarms during upset conditions?		

Facility Modifications	Response	Recommendations
36. Have personnel been relocated such that responding to equipment deficiencies (e.g., shutting down a pump with a leaking seal) may take a significantly longer time?		

Process Modifications	Response	Recommendations
1. Was a management of change system implemented before or in conjunction with the completion of the previous PHA?		
2. Is safety and health assessment or hazard evaluation documentation available for process modifications made since the previous PHA?		
3. Have process modifications been evaluated to determine whether additional engineering or administrative controls are necessary to maintain continued safe operation of the process?		
4. Have process modifications been evaluated to determine whether additional emergency response capabilities or activities are necessary?		
5. Have process modifications been evaluated to determine whether engineering or administrative safeguards have been rendered ineffective (relief capacities exceeded due to higher unit throughputs, etc.)?		
6. Have all PHA issues (e.g., siting, human factors) been addressed for equipment added to the process?		
7. Have any new utilities been directly connected to the process?		
8. Have safety system connections (e.g., nitrogen purge) been removed from process equipment?		
9. Have equipment relief valves been changed from atmospheric discharge to a closed system (or vice versa)?		
10. Are new hazards present due to new operating modes?		
11. Are new hazards present due to new process chemistry?		

Process Modifications	Response	Recommendations
12. Have the consequences of a release changed due to significantly modified operating conditions (e.g., temperature, pressure)?		
13. Have the hazards associated with equipment that has been returned to service since the previous PHA been evaluated?		
14. Have the hazards associated with equipment that has been removed from service since the previous PHA been evaluated?		

Appendix F

Facility and Stationary Source Siting Checklist

Item No.	Question	Response	Recommendations
I. Spacing Between Process Components			
1	Are operating units and the equipment within units spaced to minimize potential damage from fires or explosions in adjacent areas?		
2	Are there safe exit routes from each unit?		
3	Has equipment been adequately spaced and located to safely permit anticipated maintenance (e.g., pulling heat exchanger bundles, dumping catalyst, lifting with cranes) and hot work?		
4	Are vessels containing highly hazardous chemicals located sufficiently far apart? If not, what hazards are introduced?		
5	Is there adequate access for emergency vehicles (e.g., fire trucks)?		
6	Can adjacent equipment or facilities withstand the overpressure generated by potential explosions?		
7	Can adjacent equipment and facilities (e.g., support structures) withstand flame impingement or radiant heat exposures?		
8	When provisions have been made for relieving explosions in process components, are the vents directed away from personnel and equipment locations?		

Item No.	Question	Response	Recommendations
II. Location of Large Inventories			
1	Are large inventories of highly hazardous chemicals located away from the process area?		
2	Is temporary storage provided for raw materials and for finished products at appropriate locations?		
3	Are the inventories of highly hazardous chemicals held to a minimum?		
4	Where applicable, are reflux tanks, surge drums, and rundown tanks located in a way that avoids the concentration of large volumes of highly hazardous chemicals in any one area?		
5	Where applicable, has special consideration been given to the storage and transportation of explosives?		
6	Have the following been considered in the location of material handling areas:		
	• fire hazards?		
	• location relative to important buildings and off-site exposures?		
	• safety devices (e.g., sprinklers)?		
	• slope and drainage of the area?		
III. Location of the Motor Control Center			
1	Is the motor control center (MCC) located so that it is easily accessible to operators?		
2	Are circuit breakers easy to identify?		
3	Can operators safely open circuit breakers? Have they been trained?		
4	Is the MCC designed such that it could not be an ignition source? Are the doors always closed? Is a no-smoking policy strictly enforced?		
5	Is the MCC designed and meant to be a safe haven?		

Item No.	Question	Response	Recommendations
IV. Location and Construction of the Control Room(s)			
1	Is the control room built to satisfy current company overpressure and safe-haven Standards?		
2	Does the construction basis for the control room satisfy acceptable criteria (e.g., the Factory Mutual recommendations)?		
3	Are workers in the control room (or escape routes from the control room) protected from all of the following:		
	• toxic, corrosive, or flammable sprays, fumes, mists, or vapors?		
	• thermal radiation from fires (including flares)?		
	• overpressure and projectiles from explosions?		
	• contamination from spills or runoff?		
	• noise?		
	• contamination of utilities (e.g., breathing air)?		
	• transport of hazardous materials from other sites?		
	• possibility of long-term exposure to low concentrations of process material?		
	• odors?		
	• impacts (e.g., from a forklift)?		
	• flooding (e.g., ruptured storage tank)?		
4	Are vessels containing highly hazardous chemicals located sufficiently far from control rooms?		
5	Were the following characteristics considered when the control room location was determined:		
	• types of construction of the room?		
	• types/quantities of materials?		
	• direction and velocity of prevailing winds?		
	• types of reactions and processes?		
	• operating pressures and temperatures?		
	• fire protection?		
	• drainage?		

Item No.	Question	Response	Recommendations
6	If windows are installed, has proper consideration been given to glazing hazards?		
7	Is at least one exit located in a direction away from the process area? Do exit doors open outward? Are emergency exits provided for multistory control buildings?		
8	Are the ends of horizontal vessels facing away from control rooms?		
9	Are critical pieces of equipment in the control room well protected? Is adequate barricading provided for the control room?		
10	Are open pits, trenches, or other pockets where inert, toxic, or flammable vapors could collect located away from control buildings or equipment handling flammable fluids?		
11	Where piping, wiring, and conduit enter the building, is the building sealed at the point of entry? Have other potential leakage points into the building been adequately sealed?		
12	Is the control room located a sufficient distance from sources of excessive vibration?		
13	Is a positive pressure maintained in control rooms located in hazardous areas?		
14	Could any structures fall on the control room in the event of an accident?		
15	Is the roof of the control room free from heavy equipment and machinery?		
V. Location of Machine Shops, Welding Shops, Electrical Substations, Roads, Rail Spurs, and Other Likely Ignition Sources			
1	Are likely ignition sources (e.g., maintenance shops, roads, rail spurs) located away from release points for volatile substances (both liquid and vapor)?		
2	Are process sewers located away from likely sources of ignition?		

Item No.	Question	Response	Recommendations
3	Are all vessels containing highly hazardous chemicals or components containing material above its flash point located away from likely sources of ignition?		
4	Are the flare and fired heater systems located to minimize hazards to personnel and equipment, with consideration given to normal wind direction and wind velocity, as well as heat potential?		
VI. Location of Engineering, Lab, Administration, or Other Occupied Buildings			
1	Have site buildings been screened to identify occupied buildings requiring further review?		
2	Are occupied buildings located away from inventories of highly hazardous chemicals?		
3	Are occupied buildings located away from release points for highly hazardous chemicals?		
4	Are workers in occupied buildings protected from all of the following:		
	• toxic, corrosive, or flammable sprays, fumes, mists, or vapors?		
	• thermal radiation from fires (including flares)?		
	• overpressure and projectiles from explosions?		
	• discharges from pressure relief vents?		
	• contamination of utilities (e.g., water)?		
	• contamination from spills or runoff?		
	• noise?		
	• transport of hazardous materials from other sites?		
	• flooding (e.g., ruptured storage tank)?		
VII. Unit Layout			
1	Are large inventories or release points for highly hazardous chemicals located away from vehicular traffic within the plant?		
2	Could specific siting hazards be posed to the site from credible external forces such as high winds, earthquakes and other earth movement, utility failure from outside sources, flooding, natural fires, and fog?		

Item No.	Question	Response	Recommendations
3	Is there adequate access for emergency vehicles (e.g., fire trucks)? Are access roads free of the possibility of being blocked by trains, highway congestion, spotting of rail cars, etc.?		
4	Are access roads well engineered to avoid sharp curves? Are traffic signs provided?		
5	Is vehicular traffic appropriately restricted from areas where pedestrians could be injured or equipment damaged?		
6	Are cooling towers located such that fog that is generated by them will not be a hazard?		
7	Are the ends of horizontal vessels facing away from personnel areas?		
8	Is hydrocarbon-handling equipment located outdoors?		
9	Are pipe bridges located such that they are not over equipment, including control rooms and administration buildings?		
10	Is piping design adequate to withstand potential liquid load?		
VIII. Location of the Unit Relative to On-site and Off-site Surroundings			
1	Is a system in place to notify neighboring units, facilities, and residents if a release occurs?		
2	Are the detection systems and/or alarms in place to assist in warning neighboring units, facilities, and residents if a release occurs?		
3	Do neighbors (including units, facilities, and residents) know how to respond when notified of a release? Do they know how to shelter-in-place and when to evacuate?		
4	Are large inventories or release points for highly hazardous chemicals located away from publicly accessible roads?		
5	Is the unit, or can the unit be, located to minimize the need for off-site or intrasite transportation of hazardous materials?		

Item No.	Question	Response	Recommendations
6	Are workers in this unit protected from the effects of adjacent units or facilities for all of the following (and vice versa), and are environmental receptors and the public also protected from the following?		
	• releases of highly hazardous chemicals?		
	• toxic, corrosive, or flammable sprays, fumes, mists, or vapors?		
	• thermal radiation from fires (including flares)?		
	• overpressure from explosions?		
	• contamination from spills or runoff?		
	• odors?		
	• noise?		
	• contamination of utilities (e.g., sewers)?		
	• transport of hazardous materials from other sites?		
	• impacts (e.g., airplane crashes, derailments)?		
	• flooding (e.g., ruptured storage tank)?		
IX. Location of Firewater Mains and Backup (e.g., Diesel) Pumps			
1	Are firewater mains easily accessible?		
2	Are firewater mains and pumps protected from overpressure and blast debris impact?		
3	Is an adequate supply of water available for firefighting?		
4	Are the firehouse doors pointed away from the process area so the doors will not be damaged by an explosion overpressure?		
X. Location and Adequacy of Drains, Spill Basins, Dikes, and Sewers			
1	Are spill containments sloped away from process inventories and potential sources of fire?		
2	Have precautions been taken to avoid open ditches, pits, sumps, or pockets where inert, toxic, or flammable vapors could collect?		
3	Are process sewers that transport hydrocarbon closed systems?		

Item No.	Question	Response	Recommendations
4	Are concrete bulkheads, barricades, or berms installed to protect personnel and adjacent equipment from explosion and/or fire hazards?		
5	Are vehicle barriers installed to prevent impact to critical equipment adjacent to high traffic areas?		
6	Do drains empty to areas where material cannot pool?		
7	Can dikes hold the capacity of the largest tank?		
8	Is there a means of access in and out of dikes, pits, etc.?		
XI. Location of Emergency Stations (Showers, Respirators, Personal Protective Equipment, etc.)			
1	Are emergency stations easily accessible?		
2	Are first aid stations prudently located and adequately equipped?		
3	Are safety showers heated/freeze protected/wind protected?		
4	Is there a control room alarm for water flow to safety showers and eyewash stations? (Is there a need for such an alarm?)		
XII. Electrical Classification			
1	Is there an electrical classification document?		
2	Does the electrical classification appear correct and complete?		
3	Has the electrical classification document been recently revised?		
4	Have significant changes made since the system was originally constructed (addition of new materials, new sources of flammable gases or vapors, new low points [e.g., sumps or trenches] at grade) been included in the electrical classification document?		
5	Are the design and maintenance of ventilation systems adequate?		
6	Are there safeguards to alert operators when a ventilation system fails?		

Item No.	Question	Response	Recommendations
7	Are ventilation systems being properly maintained, and are alarms and interlocks on these systems periodically function checked?		
8	Is adequate maintenance being done to function check natural ventilation systems?		
9	Are there technical bases for design changes to the ventilation systems?		
10	Are ventilation systems verified to be adequate for new gas or vapor loads?		
11	Are there adequate controls to ensure that electrically qualified equipment is replaced with equipment of equal or higher classification?		
12	Are boundaries between electrically classified areas physical boundaries?		
13	Are Division 1 areas necessary (if there are any)?		
14	Are there adequate controls (e.g., a hot work permit system) on repair and construction activities, including work by contractor personnel?		
15	Does the electrical classification adequately reflect the effects of different modes of operation (e.g., normal operation, maintenance, startup, infrequent operating modes such as reactor regeneration or operation with a portion of the system bypassed)?		
XIII. Contingency Planning			
1	What expansion or modification plans are there for the facility?		
2	Can the unit be built and maintained without lifting heavy items over operating equipment and piping?		
3	Are calculations, charts, and other documents available that verify siting has been considered in the layout of the unit? Do these siting documents show that consideration has been given to:		
	• normal direction and velocity of wind?		
	• atmospheric dispersion of gases and vapors?		
	• estimated radiant heat intensity that might exist during a fire?		
	• estimated explosion overpressure?		

Item No.	Question	Response	Recommendations
4	Are appropriate security safeguards in place (e.g., fences, guard stations)?		
5	Are gates located away from the public roadway so that the largest trucks can move completely off the roadway while waiting for the gates to be opened?		
6	Where applicable, are safeguards in place to protect high structures against low-flying aircraft?		
7	Are adequate safeguards in place to protect employees against exposure to excessive noise, considering the cumulative effect of equipment items located close together?		
8	Is adequate emergency lighting provided? Is there adequate redundant backup power for emergency lighting?		
9	Are procedures in place to restrict nonessential or untrained personnel from entering areas deemed hazardous?		
10	Are indoor safety control systems (e.g., sprinklers, fire walls) provided in buildings where personnel will frequently be located, such as control rooms and administrative buildings?		
11	Are evacuation plans (from buildings, units, etc.) adequate and accessible to personnel?		
12	Are evacuation drills routinely conducted?		

Appendix G

Human Factors Checklist*

Item No.	Question	Response	Recommendations
I. Housekeeping and General Work Environment			
1	Are adequate signs posted near maintenance, cleanup, or staging areas to warn workers of special or unique hazards associated with the areas?		
2	Are adequate barriers erected to limit access to maintenance, cleanup, or staging areas?		
3	Are working areas generally clean?		
4	Are provisions in place to limit the time that a worker spends in an extremely hot or cold area?		
5	Is noise maintained at a tolerable level?		
6	Are alarms audible above background noise both inside the control room and in the process area?		
7	Is normal and emergency lighting sufficient for all area operations?		
8	Is there adequate backup power for emergency lighting?		
9	Is the general environment conducive to safe job performance?		
II. Accessibility/Availability of Controls and Equipment			
1	Are adequate supplies of protective gear readily available for routine and emergency use?		

* This checklist is provided for illustrative purposes only. Readers may wish to develop such a checklist specific to their own situation and needs.

Item No.	Question	Response	Recommendations
2	Are workers able to perform both routine and emergency tasks safely while wearing protective equipment?		
3	Is emergency equipment accessible without presenting further hazards to personnel?		
4	Is communications equipment adequate and easily accessible?		
5	Would others quickly know if a worker were incapacitated in a process area?		
6	Are the right tools (including special tools) available and used when needed?		
7	Is the workplace arranged so that workers can maintain a good working posture while performing necessary movements to conduct routine tasks?		
8	Is access to all controls adequate?		
9	Can operators/maintenance workers safely perform all required routine/ emergency actions, considering the physical arrangement of equipment (e.g., access to equipment, or proximity of tasks to rotating equipment, hot surfaces, hazardous discharge points)?		
10	Are valves that require urgent manual adjustments (e.g., emergency shutdown) easily identifiable and readily accessible?		
III. Labeling			
1	Is all important equipment (vessels, pipes, valves, instruments, controls, etc.) legibly, accurately, and unambiguously labeled?		
2	Does the labeling program include components (e.g., small valves) that are mentioned in the procedures even if they are not assigned an equipment number?		
3	Has responsibility for maintaining and updating labels been assigned?		
4	Are emergency exit and response signs (including wind socks) adequately visible and easily understood?		

Item No.	Question	Response	Recommendations
5	Are signs that warn workers of hazardous materials or conditions adequately visible and easily understood?		
IV. Feedback/Displays			
1	Is adequate information about normal and upset process conditions clearly displayed in the control room?		
2	Are the controls and displays arranged logically to match operators' expectations?		
3	Are the displays adequately visible from all relevant working positions?		
4	Do separate displays present similar information in a consistent manner?		
5	Are automatic safety features provided when a process upset requires rapid response?		
6	Are automatic safety features provided when a process upset may be difficult to diagnose due to complicated processing of various information?		
7	Are the alarms displayed by priority?		
8	Are critical safety alarms easily distinguishable from control alarms?		
9	Is an alarm summary permanently on display?		
10	Are nuisance alarms corrected and redundant alarms eliminated as soon as practical to help prevent complacency toward alarms?		
11	Have charts, tables, or graphs been provided (or programmed into the computer) to reduce the need for operators to perform calculations as part of the operation?		
12	If operators are required to perform calculations, are critical calculations independently checked?		
13	Does the computer check that values entered by operators are within a valid range?		
14	Do the displays provide an adequate view of the entire process as well as essential details of individual systems?		

Item No.	Question	Response	Recommendations
15	Do the displays give adequate feedback for all operational actions?		
16	Are instruments, displays, and controls promptly repaired after a malfunction?		
17	Do administrative features exist that govern when instruments, displays, or controls are deliberately disabled or bypassed and that govern their return to normal service at the appropriate time?		
18	Does a formal mechanism exist for correcting human factors deficiencies identified by the operators (e.g., modifications to the displays, controls, or equipment to better meet operators needs)?		
V. Controls			
1	Is the layout of the consoles logical, consistent, and effective?		
2	Are the controls distinguishable, accessible, and easy to use?		
3	Do all controls meet Standard expectations (color, direction of movement, etc.)?		
4	Do the control panel layouts reflect the functional aspects of the process or equipment?		
5	Does the control arrangement logically follow the normal sequence of operation?		
6	Can operators safely intervene in computer-controlled processes?		
7	Can process variables be adequately controlled with the existing equipment?		
8	Do operators believe that the control logic and interlocks are adequate?		
9	Does a dedicated emergency shutdown panel exist? If so, is it in an appropriate location?		
VI. Workload and Stress Factors			
1	Is the control room always occupied (i.e., assigned duties do not require the control room operator to be absent from the control room)?		

Item No.	Question	Response	Recommendations
2	Are the number and frequency of manual adjustments required during normal and emergency operations limited so that operators can make the adjustments without a significant chance of mistakes as a result of overwork or stress?		
3	Is the number of manual adjustments during normal operations sufficient to avoid mistakes as a result of boredom?		
4	Have the effects of shift duration and rotation been considered in establishing workloads?		
5	Is the number of extra hours an operator must work if his or her relief fails to show up sufficiently limited so that worker safety is not adversely affected?		
6	Is the number of hours an operator or maintenance worker must work during startup or turnarounds sufficiently limited so that worker safety is not adversely affected?		
7	Can additional operators (e.g., from other areas or from off site) be called in quickly to help during an emergency?		
8	Is the staffing level appropriate for all modes of operation (normal, emergency, etc.)?		
VII. Procedures			
1	Do written procedures exist for all operating phases (i.e., normal operations, temporary operations, emergency shutdown, emergency operation, normal shutdown, and startup following a turnaround or after an emergency shutdown)?		
2	Are safe operating limits documented, providing consequences of deviating from limits and actions to take when deviations occur?		
3	Are procedures current (i.e., are they revised when changes occur)?		
4	Do operators believe that the procedure format and language are easy to follow and understand?		
5	Are the procedures accurate (i.e., do they reflect the way in which the work is actually performed)?		

Item No.	Question	Response	Recommendations
6	Is responsibility assigned for updating the procedures, distributing revisions of the procedures, and ensuring that workers are using current revisions of the procedures?		
7	Are temporary notes or instructions incorporated into revisions of written operating procedures as soon as practical?		
8	Do procedures address the personal protective equipment required when performing routine and/or nonroutine tasks?		
VIII. Training (Employees and Contractors)			
1	Are new employees trained in the hazards of the processes?		
2	Do operators and maintenance workers receive adequate training in safely performing their assigned tasks before they are allowed to work without direct supervision?		
3	Does operator and maintenance worker training include training in appropriate emergency response?		
4	Do operators practice emergency response while wearing emergency protective equipment?		
5	Do operators practice emergency response during extreme environmental conditions (e.g., at night or when it is very cold)?		
6	Are periodic emergency drills conducted?		
7	Are emergency drills witnessed by observers and critiqued?		
8	Does a periodic refresher training program exist?		
9	Is special or refresher training provided in preparation for an infrequently performed operation?		
10	When changes are made, are workers trained in the new operation, including an explanation of why the change was made and how worker safety can be affected by the change?		

Item No.	Question	Response	Recommendations
11	Are operators and maintenance workers trained to request assistance when they believe they need it to safely perform a task?		
12	Are operators and maintenance workers trained to report near misses as part of the incident investigation program?		
13	Are operators trained to shut down the process when in doubt about whether it can continue to operate safely?		

Resources

1. Center for Chemical Process Safety, *Guidelines for Preventing Human Error in Process Safety*, New York: American institute of Chemical Engineers, 1994.
2. D. K. Lorenzo, *A Manager's Guide to Reducing Human Errors*, Washington: Chemical Manufacturers Association, 1990.

Bibliography

ACC 1990. *Responsible Care—A Resource Guide for the Process Safety Code of Management Practices.* Chemical Manufacturers Association, Washington, DC.

API 1990. *Management of Process Hazards—American Petroleum Institute Recommended Practice 750.* American Petroleum Institute, Washington, DC.

Burk, A.F. 1992. "Managing Change—A Multifaceted Approach," presented at the 1992 Process Plant Safety Symposium, AIChE South Texas Section.

CCPS G-1. 1992. *Guidelines for Technical Management of Chemical Process Safety.* American Institute of Chemical Engineers, Center for Chemical Process Safety, New York, NY.

CCPS G-10. 1992. *Plant Guidelines for Hazard Evaluation Procedures, Second Edition, with Worked Examples.* American Institute of Chemical Engineers, Center for Chemical Process Safety, New York.

CCPS G-27. 1995. *Guidelines for Process Safety Documentation.* American Institute of Chemical Engineers, Center for Chemical Process Safety, New York, NY.

CCPS G-62. 1999. *Guidelines for Process Safety in Batch Reaction Systems.* American Institute of Chemical Engineers, Center for Chemical Process Safety, New York, NY.

Crumpler, D. K. and Whittle, David K. 1996. "How to Effectively Revalidate PHAs," *Hydrocarbon Processing*, October 1996, pp. 55–60.

EPA 1996. *Accidental Release Prevention Requirements,* 40 CFR Part 68. Environmental Protection Agency. *Federal Register,* Vol. 61, No. 120, Washington, DC.

Frank, W. L., J. E. Giffin, and D. C. Hendershot. 1993. "Team Makeup...An Essential Element of a Successful Process Hazard Analysis," presented at the International Process Safety Management Conference and Workshop, San Francisco, CA, September 1993.

OSHA 1992. *Process Safety Management of Highly Hazardous Chemicals,* 29 CFR 1910.119. Occupational Safety and Health Administration. *Federal Register*, Vol. 57, No. 36, Washington, DC.

Smith, Kevin and Whittle, David K. 2000. "Six Steps for Effectively Updating and Revalidating PHAs," American Institute of Chemical Engineers, New York, NY. In press.

Index

R

S

T

W

Publications Available from the CENTER FOR CHEMICAL PROCESS SAFETY of the AMERICAN INSTITUTE OF CHEMICAL ENGINEERS 3 Park Avenue, New York, NY 10016-5991

CCPS Guidelines Series

Guidelines for Process Safety in Outsourced Manufacturing Operations
Guidelines for Process Safety in Batch Reaction Systems
Guidelines for Chemical Process Quantitative Risk Analysis, Second Edition
Guidelines for Consequence Analysis of Chemical Releases
Guidelines for Pressure Relief and Effluent Handling Systems
Guidelines for Design Solutions for Process Equipment Failures
Guidelines for Safe Warehousing of Chemicals
Guidelines for Postrelease Mitigation in the Chemical Process Industry
Guidelines for Integrating Process Safety Management, Environment, Safety, Health, and Quality
Guidelines for Use of Vapor Cloud Dispersion Models, Second Edition
Guidelines for Evaluating Process Plant Buildings for External Explosions and Fires
Guidelines for Writing Effective Operations and Maintenance Procedures
Guidelines for Chemical Transportation Risk Analysis
Guidelines for Safe Storage and Handling of Reactive Materials
Guidelines for Technical Planning for On-Site Emergencies
Guidelines for Process Safety Documentation
Guidelines for Safe Process Operations and Maintenance
Guidelines for Process Safety Fundamentals in General Plant Operations
Guidelines for Chemical Reactivity Evaluation and Application to Process Design
Tools for Making Acute Risk Decisions with Chemical Process Safety Applications
Guidelines for Preventing Human Error in Process Safety
Guidelines for Evaluating the Characteristics of Vapor Cloud Explosions, Flash Fires, and BLEVEs
Guidelines for Implementing Process Safety Management Systems
Guidelines for Safe Automation of Chemical Processes
Guidelines for Engineering Design for Process Safety
Guidelines for Auditing Process Safety Management Systems
Guidelines for Investigating Chemical Process Incidents
Guidelines for Hazard Evaluation Procedures, Second Edition with Worked Examples
Plant Guidelines for Technical Management of Chemical Process Safety, Revised Edition
Guidelines for Technical Management of Chemical Process Safety

Guidelines for Process Equipment Reliability Data with Data Tables
Guidelines for Safe Storage and Handling of High Toxic Hazard Materials
Guidelines for Vapor Release Mitigation

CCPS Concepts Series

Revalidating Process Hazard Analyses
Electrostatic Ignitions of Fires and Explosions
Evaluating Process Safety in the Chemical Industry
Avoiding Static Ignition Hazards in Chemical Operations
Estimating the Flammable Mass of a Vapor Cloud
RELEASE: A Model with Data to Predict Aerosol Rainout in Accidental Releases
Practical Compliance with the EPA Risk Management Program
Local Emergency Planning Committee Guidebook: Understanding the EPA Risk Management Program Rule
Inherently Safer Chemical Processes: A Life-Cycle Approach
Contractor and Client Relations to Assure Process Safety
Understanding Atmospheric Dispersion of Accidental Releases
Expert Systems in Process Safety
Concentration Fluctuations and Averaging Time in Vapor Clouds

Proceedings and Other Publications

Center for Chemical Process Safety International Conference and Workshop: Process Industry Incidents—Investigation Protocols, Case Histories, Lessons Learned, 2000
Proceedings of the International Conference and Workshop on Modeling the Consequences of Accidental Releases of Hazardous Materials, 1999
Proceedings of the International Conference and Workshop on Reliability and Risk Management, 1998
Proceedings of the International Conference and Workshop on Risk Analysis in Process Safety, 1997
Proceedings of the International Conference and Workshop on Process Safety Management and Inherently Safer Processes, 1996
Proceedings of the International Conference and Workshop on Modeling and Mitigating the Consequences of Accidental Releases of Hazardous Materials, 1995
Proceedings of the International Symposium and Workshop on Safe Chemical Process Automation, 1994
Proceedings of the International Process Safety Management Conference and Workshop, 1993
Proceedings of the International Conference on Hazard Identification and Risk Analysis, Human Factors, and Human Reliability in Process Safety, 1992
Proceedings of the International Conference and Workshop on Modeling and Mitigating the Consequences of Accidental Releases of Hazardous Materials, 1991
Safety, Health and Loss Prevention in Chemical Processes: Problems for Undergraduate Engineering Curricula